Ways of Knowing

Science and Mysticism Today

Edited by Chris Clarke

IMPRINT ACADEMIC

Published in the UK by Imprint Academic
PO Box 200, Exeter EX5 5YX, UK

Published in the USA by Imprint Academic
Philosophy Documentation Center
PO Box 7147, Charlottesville, VA 22906-7147, USA

ISBN 1 84540 012 7

A CIP catalogue record for this book is available from the British Library and US Library of Congress

OF RELATED INTEREST FROM IMPRINT ACADEMIC

Science, Consciousness and Ultimate Reality
ed. David Lorimer

Grassroots Spirituality
by Robert KC Forman

Cognitive Models and Spiritual Maps
ed. Jensine Andresen & Robert KC Forman

The Varieties of Religious Experience: Centenary Essays
ed. Michel Ferrari

God In Us: A Case for Christian Humanism
by Anthony Freeman

Shadow, Self, Spirit: Essays in Transpersonal Psychology
by Michael Daniels

Contents

Acknowledgments   iv
About the authors   v
Introduction   1

Section 1: The Social Context   5

Subjugated Ways of Knowing   *June Boyce-Tillman*   8
Creativity as the Immune System of the Mind
  and the Source of the Mythic   *John Holt*   34
Soul's Sanctuary:
  Mystical Experience as a Way of Knowing   *Jennifer Elam*   51

Section 2: The Perspective of Psychology   67

Attachment Mechanisms and the Bridging of
  Science and Religion   *Douglas Watt*   70
There Is a Crack In Everything:
  That's How the Light Gets In   *Isabel Clarke*   90

Section 3: Physics, Logic and the Pluralistic Universe   101

Spiritual Knowing:
  A Participatory Understanding   *Jorge N Ferrer*   107
Ignacio Matte Blanco and the Logic of God   *Rodney Bomford*   129
Both/And Thinking: Physics and Reality   *Chris Clarke*   143
Ways of Knowing and the Quest for Integration   *Lyn Andrews*   159

Section 4: The Nature of the Spiritual Path   177

Ordinary and Extra-ordinary Ways of Knowing
  in Islamic Mysticism   *Neil Douglas-Klotz*   180
Earth in Eclipse   *David Abram*   200
Ecological Awareness:
  A Meeting of Science and Mysticism   *Anne Primavesi*   218

Final reflections   234
References   237
Index   247

Acknowledgments

The diagram on p. 11 by Carter Heyward is reproduced with her permission.

Extracts from Jorge Ferrer's *Revisioning Transpersonal Theory: A Participatory Vision of Human Spirituality* appearing in his chapter are reproduced by kind permission of the State University of New York Press.

The chapter by David Abram is adapted from a version previously published in *Tikun*.

Note on Cover Image

The image of the sun and moon joined, incorporated into the cover design, comes from alchemical symbolism, where it represents the integrative culmination of the alchemical Work (see Jung [1968], pp. 75, 241, 244). A variety of integrations are named in this book: of the dominant (male) culture with the subjugated (female) culture; of the emotional and intellectual/relational and propositional sides of the mind and brain; of the heights and the depths of human experiencing; of the Maximum and the Minimum. Readers are invited to find their own interpretation of this particular rendering of the image.

About the Authors

June Boyce-Tillman is Professor of Applied Music at University College Winchester. She pioneered work in introducing composing activities into the classroom and has a particular interest in Music and Theology including Religious Education. She regularly writes and takes workshops linking these areas together. She has done pioneering work in interfaith dialogue, writing articles and speaking on interfaith and intercultural links in Britain and abroad. Her most recent publications are in the areas of music, healing and spirituality and the medieval abbess, Hildegard of Bingen.

John Holt was a lecturer in the School of Fine Art, Art History and Cultural Analysis at Leeds University and then Fellow in Art and Design at Loughborough University. An artist and cultural activist, he has written, from practical experience, on Native American and Aboriginal culture, the arts of South Asia and the status of those defined as 'mentally ill'. His work on the possibility of transformation through creativity led to him founding AIM (Artists in Mind), a charitable organisation set up to promote and explore creativity in those in emotional and spiritual crisis.

Jennifer Elam is a licensed psychologist who has taught at the college level, worked in residential treatment, and worked in schools with students aged preschool through adult. As a Cadbury scholar at Pendle Hill she listened to many people's stories of their experiences of God and recorded about one hundred of them. many of which came to influence the paintings that she was creating. She presently leads art retreats, facilitates programs at the Listening Center in Springfield, Pennsylvania, works as a psychologist, and makes time to write and paint. Her heart's desire now is to enjoy ordinary life.

Douglas Watt has been a clinical neuropsychologist for roughly 18 years after graduating from Boston College and Harvard University for his PhD and BA. He has directed Psychology and Neuropsychology departments in two teaching hospitals in the Boston area and is currently Instructor in Neuropsychology, Boston University School of Medicine. He has had a passionate long-term interest in virtually any and all perspectives on emotion, and believes that only through inter-disciplinary work that any real progress will be made in clarifying the deep mandates of emotion as part of our evolutionary heritage.

Isabel Clarke is a Consultant Clinical Psychologist, currently working for the Hampshire Partnership NHS Trust, providing a psychological therapies service for an inpatient psychiatric hospital near Southampton. She is a lifelong practising Anglican, and active in the Association for Creation Spirituality (Greenspirit). She has published and given work-shops and lectures on the interface between psychosis and spirituality since 1999, including the edited book *Psychosis and Spirituality: Exploring the New Frontier* (Whurr, 2001).

Lyn D Andrews is a secondary school science and biology teacher who became aware of an inner calling to start writing fiction in 1994. This led to a renewed interest in the relationship between science and religion, culminating in a life-changing mystical experience in 1996. Since then she has concentrated her efforts in gaining scientific support for an interconnected, creative view of the universe. She believes passionately that increasing self-awareness and self-acceptance leads to a more enriched and fulfilling life and that ultimate co-creative power resides within us. She is currently in the process of establishing a new approach to education and healing called Eduspirit.

Jorge N Ferrer is Associate Professor at the California Institute of Inte-gral Studies, San Francisco, and Adjunct Faculty at the Institute of Transpersonal Psychology, Palo Alto. Formerly a fellow of 'La Caixa' Foundation, a research fellow of the Catalonian Council, and an ERAS-MUS scholar at the University of Wales (United Kingdom), his writing includes *Revisioning Transpersonal Theory: A Participatory Vision of Human Spirituality* (SUNY Press, 2002). In 2000, he received the Presi-dential Award from the Fetzer Institute for his seminal work on con-sciousness studies.

Rodney Bomford studied Mathematics at Oxford and subsequently theology at Oxford, the College of the Resurrection, Mirfield, and Union Seminary, New York, specialising in Philosophy of Religion. He was ordained in the Church of England and from 1977 to 2001 was Vicar of St Giles' church, Camberwell. He was a founding member of the Lon-don Bi-logic group which for nearly 20 years has pursued the thinking of the psycho-analyst Ignacio Matte Blanco and is now part of an inter-national network. In his book *The Symmetry of God* (1999) he explored the connections between God and the Unconscious in the light of Matte Blanco's theories.

Chris Clarke was Professor of Applied Mathematics and Dean of the Faculty of Mathematics at the University of Southampton, where he is now a Visiting Professor. He has published three books on General Relativity and papers on relativity, astrophysics, cosmology, the foundations of quantum theory, biomagnetic imaging, the physics of consciousness and ecotheology.

Neil Douglas-Klotz is co-chair of the Mysticism Group of the American Academy of Religion and directs the Edinburgh Institute for Advanced Learning in Edinburgh, Scotland. He is an independent scholar of religious studies, spirituality, and psychology, and author of many books in this area including *Prayers of the Cosmos* (1990), *Desert Wisdom* (1995) and *The Sufi Book of Life* (2005). He holds a PhD in religious studies and psychology from Union Institute University and taught these subjects for ten years at Holy Names College in California. He has followed the practices of the Sufi path since 1976 and was recognized as a senior teacher (murshid) in this tradition in 1993.

David Abram, cultural ecologist and philosopher, is the author of *The Spell of the Sensuous: Perception and Language in a More-than-Human World* (Vintage, 1997), for which he received, among other awards, the Lannan Literary Award for Nonfiction. He has lived with indigenous sorcerors in Indonesia, Nepal, and the Americas, and his writings have appeared in academic and other journals. He has also been named by The Utne Reader as one of a hundred leading visionaries currently transforming the world.

Anne Primavesi is a Fellow of the Centre for the Interdisciplinary Study of Religion, Birkbeck, London, and of the Westar Institute for the Advancement of Religious Literacy, Santa Rosa, California. Formerly Research Fellow in Environmental Theology, University of Bristol, her publications on theology and science include most recently *Gaia's Gift: Earth, Ourselves and God after Copernicus* (Routledge, 2003).

Chris Clarke

Introduction

What does it mean, to *know*? Consider these quotations ...

> My mother would get up early. She would go outside and stand there a long time. Then she would say, 'Vehsih yehno nah ha ooh.' That means, 'The caribou are just under the mountains over there, and they're coming.' Everyone would get excited. (Norma Kassi)[1]

> Not only do we know more about the universe, but our understanding is deeper, and the questions that we are asking are more profound. Still, our understanding of the origin and evolution of the universe has not yet caught up with what we know about it. (Wendy L Freedman)[2]

> Then in the distance I began to see ... the physical cosmos and the underlying constitutive forces that built the universe and sustain it. ... I learned by becoming what I was knowing. I discovered the universe not by knowing it from the outside but by tuning to that level in my being where I was that thing. (Chris Bache)[3]

> The sapiential perspective envisages the role of knowledge as the means of deliverance and freedom, of what the Hindu calls moksa. To know is to be delivered. (Seyyed Hossein Nasr)[4]

These are about very remarkable, and very different, ways of knowing. They seem to go beyond the knowing of our more ordinary life, which is concerned with familiarity with people and places, the ingrained ability to perform various tasks, or our accumulated learning about the consequences of our actions. The wisdom of Norma Kassi's mother, an elder of Gwich'in Nation, of Yukon, is intensely practical and born of a lifetime of living close to nature. The knowledge of the cosmologist Wendy Freedman is derived from measurements from satellite observatories orbiting the earth, coupled with the full intellectual apparatus of modern theoretical physics; it is vast but seemingly remote from our lives. The vision of Chris Bache, seen in the trance of a

[1]   Kassi (1996), p. 75.
[2]   Freedman (2003).
[3]   Bache (2000).
[4]   Nasr (1989), p. 309; quoted in Ferrer (2002), p. 127.

psychedelic state of consciousness, claims to deliver similar cosmological information, but through direct awareness with no instrument other
than the body–mind. And the knowledge dealt with by Sayed Hussein
Nasr, knowledge of the ultimate nature of all existence, is attained
through the long refinement of consciousness taught in traditional
meditative spiritual paths.

Is it right to call all these 'knowing', as if it were a question of a single
human activity applied to different areas; or are they so different that it
is misleading to use the same word for all of them? Do they fundamentally differ from the more pedestrian knowings of everyday life, or is it
more a matter of degree? What do we mean when we assess the particular claims of each as 'right' or 'wrong'?

For over a thousand years, and in many cultures, attempts have been
made to answer these questions by appeal to a hierarchy of ways of
knowing — an ascending chain of types of knowledge, each superior to
the one below. At different times, science or religion have each claimed
the pinnacle of knowing, the knowing at the top of hierarchy in terms of
which everything else, whether theoretical or practical, could be
derived. A famous modern example of this on the scientific side is Francis Crick's *Astonishing Hypothesis* that the whole of life and mind can be
explained in terms of biochemistry and the interactions of neurons. An
alternative claimant on the spiritual side might be Ken Wilber whose
collection of writings[5] gives a pinnacle place to the sort of spiritual
knowing being described by Nasr. Both these examples have come in
for trenchant criticism, as well as enthusiastic praise. In the light of the
new discoveries surveyed in this book, it now seems necessary to
explore ways of knowing in which there is no boss-knowledge, no
supreme ruler at the pinnacle.

Our aim in this book, therefore, is to consider the possibility that
many ways of knowing need to be recognised alongside each other,
without a hierarchical structure of superiority one to another, to examine different ways in which this can be so, and explore the consequences
of this for how we might live our lives. There is a need to proceed both
boldly and skilfully. Within systems that have an order of superiority
between knowings there is a vital distinction between those where the
higher ways negate and replace the lower, and those where the higher
ways incorporate and then go beyond the lower; a distinction between
the malevolent strict hierarchies and the benevolent *holarchies,* as
Wilber terms his own system of levels that incorporate the lower ones.
Boldness is needed in order to expose the injustices that have been per-
petrated by the dogmatic wielding of hierarchical power. Skill is

---

[5]   E.g. *Sex, Ecology, and Spirituality* (1995).

needed to understand the gradations of benign and malevolent versions, and be always alert to the tendencies of benignly inclusive schemes to slide over into the camp of their authoritarian hierarchical cousins.

The chapters that follow are grouped into sections which cover the different aspects of a new vision of our knowing. In keeping with the spirit of alternative ways, academic analysis and story-telling will be found side by side. Each section begins with a brief introduction in which I describe its place in the overall development of the argument of the book.

First, the **social context** will be examined, revealing the forces that have shaped the restricted way of knowing that has become 'normal' in the West, and the damage that has thereby been done to individuals, to society, and indeed to the planet. June Boyce-Tillman, whose study of music in society has led her to an analysis of the nature of society itself, categorises the ways of knowing that have been subjugated. This is followed by two chapters whose authors have worked closely with people who have often been repressed because of their way of knowing. Both identify the experiences involved as *mystical,* though they usually lie outside established religious systems (these are considered later). John Holt describes this from his experience of the role of Art in bringing about self realisation in those who have been confined to penal mental institutions. Then Jennifer Elam presents a panorama of the variety of different spontaneous experiences, continuing John Holt's account with a focus on the way in which society, through its narrow commercially driven definition of 'normal' has labelled these experiences as pathological.

The approach of science is then entered through **the perspective of psychology,** again with two complementary accounts. Douglas Watt presents the viewpoint of neuropsychiatry, giving a powerful plea for the rebuilding of the moral framework that has been eroded by the narrowness of both conventional religion and conventional scientism. Isabel Clarke then describes a cognitive approach that roots the dysfunctions of self and society, described previous chapters, in the fundamental nature of the human being, and develops a conception of knowing that is based on the divided nature of the knower, ourselves.

The next section, on **physics, logic and the pluralistic universe,** surveys the fundamental change in our world view that arises if we accept the validity of alternative ways of knowing. If we are to bring mystical and subjugated ways alongside the scientific way of knowing, then we have to make a fundamental revision in the philosophical assumption that has so underpinned science, namely *realism:* the view that the world is simply 'there', outside us, waiting to be passively observed by us (or minor variations of this). Until recently there has been no alternative to realism that has done justice to the actual nature of science. Now, however, the development of *participatory philosophies* enable us to go beyond realism. In this section, Jorge Ferrer presents his definitive version of a participatory conception of the world that affirms the fundamental place of the Mystery at the core of our experience and at the same time makes sense of the multiplicity of shores on the ocean of this mystery. I see this as the first world view that genuinely acknowledges the experiences of the different mystical traditions and of science. Then Lyn Andrews gives a detailed account of her own spontaneous mystical experience, which describes a process of progressive transformation of life and of progressive growth in understanding following an initial revelation, leading to a remarkable vision of our place in the cosmos.

Adopting this view requires us, however, to alter the basic logical structures of our thinking. This is described first by Rodney Bomford, who demonstrates the power of an enlarged system of logic that unites mystical experience with the data of psychoanalytic research. Then we move to physical science in a chapter where I show how this new logic is further extended by modern physics. I also discuss here the limits that this implies to the scope of conventional science within this larger framework. Together these chapters provide the radical conceptual and intellectual structure that is needed for the new world view and the new science emerging from the previous sections.

Mysticism is not about feelings or concepts, it is about living. And so the final section examines **the nature of the spiritual path** as it is displayed by all we have learnt. Drawing on Middle Eastern mysticism, Neil Douglas-Klotz expounds a mysticism of ordinary life — but seen in an extra-ordinary way. Next David Abram returns us to the ground of all our experience and all our living: the recovery of our intimate relationship with the planet and all its beings, human and other than human. Finally this central role of the planet is named by Anne Primavesi as Ecology, in a chapter that sets out a path that integrates science and theistic religion within an ecologically based spiritual path involving multiple ways of knowing.

I close the book with a brief reflection on the relevance of this inclusive vision to our struggling species.

Section 1

The Social Context

Examining different ways of knowing does not stay innocuously within philosophy, but takes us straight into politics. Those who have developed critiques of a hierarchical approach have drawn attention to the way in which a hierarchy of ways of knowing tends to be connected with a hierarchy of political power among classes of a society. In Europe, the battle for dominance in power between scientific knowing and the religious knowing of the church was won by the former, as the church progressively ceded more territory to Newton, Darwin, Freud and Hawking. A truce was called when the church was left only the comparatively worthless ground of official morality. In recent years, however, feminist thinkers[1] have realised that while these men of science and men of the church played their power games, the 'lower' ways of knowing continued, but in forms that were increasingly suppressed, hidden and forgotten. Below the rulers of the power/knowledge hierarchy there persisted what Foucault[2] termed 'subjugated ways of knowing', including the practical and spiritual knowing of women, until quite recently handed down orally and unrecorded in the histories written by men. The knowledge hierarchy was identified as a patriarchy, and it became clear that the collision between the subjugated women's knowing and the patriarchal/hierarchical knowing of the church had resulted in the witch trials that culminated in the sixteenth century. This was the most malevolent of all hierarchies.

From this perspective, it became clear that there was in fact nothing 'lower' about the subjugated ways. Their contribution to human well being, individual and social, it was suggested, was even greater than that of the dominant ways. And a similar pattern was played out in the case of the indigenous peoples of North America, Australia and the other lands that were conquered by the dominant power of the West, and whose cultures then contributed to the hidden wealth of the subju-

---

[1]   See, for example, Merchant (1983).
[2]   Foucault (1980), pp. 81, 84.

gated ways of knowing. Thus the study of the role of hierarchy in ways of knowing has become a means of liberating the oppressed.

In the following chapter June Boyce-Tillman explores this approach by examining ways of knowing that have not been validated by the dominant culture. She looks at the need for a dynamic balance, within the self and within the wider society, between the qualities valued by the dominant and subjugated cultures, identifying in this way polarities such as process/product, challenge/nurture, the individual/the community and the embodied/the disembodied. In the course of this, she shows how particular dominant value systems, when pushed to extremes, turn sour but how in right relationship with those value systems which are subjugated they retain their integrity. The chapter looks towards a genuinely inclusive society in which a variety of ways of knowing are valued.

The restoration of this right relationship between subjugated and dominant poles is, however, only possible when the subjugated pole is recognised *as* a way of knowing *independently and in its own right*. So long as classical science is regarded as the supreme way to which everything else can be reduced, so long as science sees intuitive, mystical, artistic and other modalities as only emotional glosses on scientific 'facts', then, while we might achieve mutual tolerance, there can be no genuine mutual interaction as equals, as is needed in order to heal our society. We will see later in the book how psychology, modern physics and a participatory philosophy now can establish this crucial mutuality of different ways.

This theme of the need for the restoration of the separated poles of knowing is continued through the next two chapters, with John Holt first taking it up from the vantage point of the visual arts. He describes how the dominant value system has cut itself off from the our mythological story-telling and from our immediate sensory connections with our living planet. Then he focuses on his experience in working with the most marginalised of our culture, the 'unforgiven' in our secure mental hospitals, who find themselves the despised and repressed carriers of the shadow form of all that the dominant culture has rejected. Here in a great many of the cases where it has been facilitated, the creative potential of the human being has flowered as an 'immune system of the mind' in response to suffering, as a means of finding wholeness for the individual, and hence as a way of pointing society to wholeness. In a challenging chapter he offers evidence that it is precisely here, where the suffering of a sick society is manifest, that the healing capacity of the repressed way of knowing can appear. He explains how creativity has a natural tendency towards a heightened sense of self-realisation in the individual, constituting a process of clarification of the relationship between self and the world — self and body, self and environment, self and God — leading him to compare the spiritual insights derived from this creativity with those of classical mystics such as John of the Cross. In this he echoes Matthew Fox's 'Creation Centered Spirituality' in which 'every mystic is an artist and every true artist is a mystic'.[3]

These themes are continued by Jennifer Elam, based on her own experience and on the stories of about one hundred people to whom she has listened in depth. Extracts from a few of these stories are presented here. As a result she is convinced that 'Mystical creative energy is not just something of the past. It is alive today. Many people continue to open themselves to those possibilities, and as a result, they begin to *know* that which is beyond daily life. They begin to experience something of the numinous that is ineffable. And more often than not, this *knowing* invites something new to be created into the world.' These stories also, however, bear witness to the extent to which these mystical ways of knowing have been repressed as a result of partitions erected both by science and by many forms of institutionalized religion. Those who are open to Spirit are labelled as 'abnormal', while the definition of 'normal' has become more and more narrow. 'This', she writes, 'has created a greater realm of deviance; pathology and criminality increases as our tolerance and acceptance of differences decreases. Intolerance and the profit motive have united in the recent past to usher in despair as the modus operandi of society. We have moved from educating children to value basic humanistic principles to educating them to be unquestioning consumers.'

As with John Holt, her experience also leads her to the hope that transformation in individuals and in society — 'overcoming the roadblocks' — is possible, provided that we understand the processes that are needed. As June Boyce-Tillman has charted in her chapter, the path to healing creativity usually passes through the chaos of de-integration. From this, Jennifer Elam finds, 'We emerge stronger, fuller, endowed with a deeper level of knowing, more able to share our knowing with others.' This is the path whereby the mystical breaks through into society; and when that happens 'destruction, violence and war become impossible'.

---

[3]   Fox (1988), p. 58.

June Boyce-Tillman

Subjugated Ways of Knowing

Introduction

My aim in this chapter is to examine the way a culture validates itself by the social construction of its value systems. I will establish a model of these value systems and look at ways of bringing the apparent polarities together.

In a growing movement in the early twenty first century the West is trying to heal a rift that has developed in its intensely rationalistic culture. Gooch[1] defines this rift in terms of two systems of thought, both of which co-exist in the human personality. The favoured characteristics of one system (System A) are

- activity leading to products
- objectivity
- impersonal logic
- thinking and thought
- detachment
- discrete categories of knowledge which is based on proof and scientific evidence.

The other system (System B) favours

- being
- subjectivity
- personal feeling
- emotion
- magic
- involvement
- associative ways of knowing
- belief and non-causal knowledge

[1]   Gooch (1972).

He suggests that the Western world has chosen to value the first of these value systems. The second has therefore become devalued. I have called the ways of knowing that characterise System B subjugated ways of knowing, a term based on such theorists as Foucault[2] and Belenky.[3]

There is an inextricable relationship between the individual and the society in which s/he lives. Theories of personality, like the inventory of the Myers-Briggs Type Indicator[4] based on the work of Jung, which classifies personalities in terms of types, show that we can identify certain individuals as Type A, that is, those who are happy acting on logic and scientific reasoning which are part of System A. Others can be called Type B and will favour system B, acting intuitively and valuing belief and magic. This classification thus serves to highlight the relationship between the individual personality and society and the roots of some human dis-eases. In a society where the individual's way of knowing is in tune with that of the society the person is more likely to be seen as well-adjusted and will suffer less stress and dis-ease. It is clear that the type A people will feel more at ease than type B in western society Type B people, on the other hand, are more likely to exhibit signs of dis-ease and, indeed, to be classified as 'abnormal' by the surrounding society. However, type A people will also have the type B capacities within themselves and these will require exploration to achieve a fully rounded humanity. Similarly type B people can use the prevailing values in the culture to develop the less favoured aspects of their personality. What is clear from the literature on the use of leisure in our society[5] is that people do use their leisure time re-balance their life of work.

What is helpful for this book is the notion that normativity is established by those in power by means of the exclusion of the deviant 'other'. These subjugated ways of knowing are always in flux and cannot be defined specifically but only in relation to the dominant value system of any particular culture at a particular time. I will illustrate this with a story of a culture whose dominant values are different from those in the West:

> I was privileged to spend some time with a native people in North America. I had been present at several sweat lodges at which a particular medicine man had been working. I had also purchased a small hand drum and he had consented to beat the bear spirit into it for me. One evening he was preparing for a sweat lodge and said that he needed a powerful woman to help him and sit alongside him. This would usually be his wife but she was unable to be there so

[2]   Ball (1990).
[3]   Belenky et al (1986).
[4]   Myers and McCaulley (1985), Myers (1993).
[5]   Wilensky (1960), pp. 32–56, and Stebbins (1992).

would I help him. I was both honoured and terrified but he said he would help me with the ritual and so I agreed to the role. The first round of prayer took place and he concluded it by saying that now June would sing a song about the eagle and the sunrise. It was here that I thought that I had met an insuperable problem. But I remembered that in the songs I had heard each phrase started high and then went lower in order to bring the energy of the sky to the earth. So I started each phrase high and took it lower, singing about the eagle and the sunrise. It was a powerful experience for me and my voice seemed to come from a place of power deep inside that I had not experienced before. With the prayer round ended we went outside to cool in the night air. 'Great song, June' said the leader of the sweat lodge. I was about to say that, of course, I had to make it up and then remembered from my previous conversations with some of the women singers that in this culture everything is given not the creation of an individual. So I replied: 'The Great Spirit gave it to me when I was in the Lodge.'

Here I was in a culture where the intuitive way of receiving material construed as coming from a connection with a spiritual source was the dominant way of knowing, not the individualistic, humanistic way of the individual composer creating an individual song from their own experience in their own personal subconscious. I had previously been in a women's sweat lodge after which the leader had said that she thought the Great Spirit had given her the whole of the song when she was in the Lodge and that previously she had only received part of it. In the West we would probably have said something like that we had not yet finished it and were working at completing it.

The Enlightenment saw a massive swing to a different value system in which the intuitive visionary experience, is at best marginalized and at worst pathologised. Western Post Enlightenment culture now sees reason as paramount and this has coloured the way in which all creative expression is regarded. Order and clarity and a desire to see the world as defined by scientific reality — 'as it really it is' — have been the ultimate goals. The chaotic, the imaginary, the obscure have been seen as enemies and indeed requiring of a cure that can be achieved largely by reason. This polarity between reason and intuition is but one example of the polarities that make up the value systems that exist within each person and within any particular culture. Feminists have linked this with dominance of masculinist views.

Although many of the quotations in this Chapter are taken from feminist literature of one discipline or another, the use of the term subjugated ways of knowing rather than women's ways of knowing acknowledges that there are a variety of subjugated groups of which women are only one. There is a clear link between those who hold power and which ways of knowing will become subjugated, defined by the value systems of those in power. Moreover, those who are dominant will put systems in place to see that they remain dominant, like prisons, mental hospitals and economic difference.

Western society has embraced the Type A set of values. But we are not concerned here with just a single polarity. Many other value systems exist which are reflected in other cultures, all of which are subjugated within western culture. Carter Heyward (2003) in this diagram shows clearly the effects of patriarchy and shows the variety of those who are likely to become subjugated knowers.

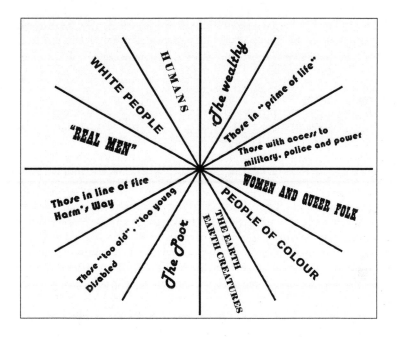

This opens up a possible link between the selfhood of the individual knower and the wider society. The 'shadow' (the subjugated ways of knowing of those in power) becomes the 'shadow' (the marginalized Other) of the wider society. Drawing on a wide variety of literatures from education, therapy and humanistic psychology[6] sociology, gender studies and theology I have identified certain polarities which are in dynamic relationship. This model identifies a number of polarities that are in dynamic relationship within the self and within society.

The model presents a way of looking at the health of a society and a person. In a balanced person and a balanced society all the possibilities

---

[6]   Particularly Dewey (1910), Maslow (1962), Lieberman (1977), Rogers (1976), Tillman (1987), Assagioli (1994), Helman (1994), Aldridge (1996).

are held as of equal value but in most people and societies this is not the case. It is a both/and logic not an either/or. This will be because of issues of power such as those identified above and the consequent establishing of a normative dominant discourse. However, the more dominant one of the pairs that make up the polarities becomes, the more the alternative value system will tend to be projected onto the marginalized 'other'; which will consist of groups who have no power. Within the self, as theorists like Myers[7] have written, each of us chooses particular ways of knowing so that the others become subjugated – making up the 'shadow' as Jung called it.

This chapter looks at these polarities and how they have functioned at various times in various Christian cultures both in the Western cultures and those that were colonized by that culture.[8]

The Model of Self and Society

These polarities are drawn as having a constant flow between them. Balance is defined as being when that flowing is fluid and dynamic. The subjugated and dominant knowings have to be in dialogue with one another to achieve a wholeness. The dominant culture will validate one of the poles more highly than the other, so effort will be required to keep the flow moving to the subjugated way of knowing. When the balance is achieved, the self or the society will have developed the new paradigm searched for by many writers.[9] In the following sections we shall examine what happens when this dynamism is broken.

As we discuss the way the subjugated ways of knowing can be brought into relationship with the dominant ways, we will look at them from the point of view of the individual as well as the structures of society, where we shall look at new paradigms of organization. Each section will start by looking at ways of balancing in the wider society and then look at the effect on the individual. In each of these we will examine both the effect of the dynamic balance becoming sluggish and the effect of the dynamic connection breaking down altogether. Those sections which are less directly relevant to the other chapters of this book will be presented in summary form.

---

[7]   Myers (1993), Myers and McCaulley (1985).
[8]   This is slightly modified from the model as set out in Boyce-Tillman (2000). I am indebted to Stuart Manins who shared his work with the Maori people in New Zealand for help in refining the model.
[9]   E.g. Donna Haraway (1992), Chela Sandoval (2000) and Chris Clarke (2002).

1 Unity/Diversity

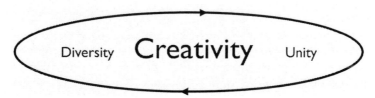

Diversity **Creativity** Unity

All descriptions of the processes of creativity include a measure of chaos or darkness — a time when the whole appears to fragment before it re-establishes itself again in a different configuration.[10] Thus the free-flowing balance between these poles is vital for a creative and inno-vative society and a vibrant and flexible self. The Cartesian view of the unified, separated self has, however, been central to the project of West-ern rationalism. It posited an 'essential' self to which various attributes like gender and race could be attached. It was attached to a notion of a steady march towards a single ideal state which was often equated with the 'light' both of the Christian Scripture and the Enlightenment project as a whole. It perpetuated the more violent aspects of social policy. The imposition of a unitary ideal on society characterizes neo-colonial enterprises: the spread of multi-nationals like McDonalds, or the enter-prises of performing groups of Western origin like orchestras and rock groups. Postmodernism has encouraged moves beyond this to the valu-ing and respect for 'difference'. The socio/political implications of the fictive unity of the self need challenging.

The degree of tolerance for diversity is an important element in the way a community defines itself — in terms of how many groups it defines as Other. The tightness of the boundaries established by a par-ticular society is a product of the degree of threat that is seen either from without or more significantly, from within. And yet it is by the admit-ting of diversity that new societies and new ways of conceiving society have emerged. When the two poles of unity and diversity are balanced in a society it can grow and change creatively. We have seen the process

[10] This process of creativity is seen (Wallas, 1926) as including various phases, which are adjusted, fine-tuned and restructured by various writers. These include preparation, the exploration of possibilities and generating of ideas, incubation, which involves less conscious activity, illumination, the 'eureka' experience and elaboration, the working out of the project in a tangible form. The incubation phase can be likened to a descent into a personal underworld, chaos or darkness. Importantly a descent into the personal unconscious or subconscious with a somewhat chaotic nature is seen as an important component of the incubation phase. I have explored this in my performance *Lunacy or the Pursuit of the Goddess* (2002).

of readjusting this balance in a post-apartheid South Africa or Australia's struggle to acknowledge the rights of its indigenous aboriginal
population. In the West we have placed a high value on unity — our
children live in a world where normalization is taking control with
standardized testing for all children regardless of race, economic class,
gender and all the factors that make children various rather than the
same. The end of the binary split is Fascism which is a perfectly unified
society where diversity has been obliterated by violence. Once the
polarities become fixed, cut off from a flowing interchange, the risk of
the supremely tolerant society is that it will slip into anarchy. The free
flowing between these polarities produces a creative and innovative
society.

The Christian Church in Europe has put a high premium on unity and
the dominance of a single metanarrative. These dominating values
have gone alongside a concern for order and the privileging of order
over chaos which has been demonized in such hymns as:

> Thou whose Almighty Word
> Chaos and darkness heard
> And took their flight.

Here we see order, light and truth inextricably linked in the images presented. Feminist theology and liberation theology have problematised
such thinking[11] following in the steps of postmodern and postcolonial
theorists like Paul Ricoeur who see such thinking as linked with all
great colonial enterprises.

To summarise this in a schema of the progressive breakdown of flow:

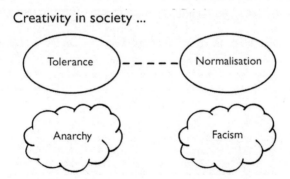

Creativity in society ...

Tolerance — — — — Normalisation

Anarchy   Facism

Moving to a personal level, the pursuit of the integrated self and the
perception of the journey as being a straight and steady progress have
resulted in a certain internal fascism, which we may describe as rigidity
or, indeed, obsession. The pattern of the creativity of the self is a cyclical

[11]   Ward and Wild (1995).

exploration of deintegration and reintegration. Because of the high premium placed on integration, deintegration is frequently pathologized. A classic example is the grieving process taking place when a loved person, animal or object disappears. The self has previously integrated with that person, animal, object in a particular place. When they disappear, the self has to deintegrate in order to reintegrate with a previous part missing. Hannah Wild and Jennifer Ward in their book entitled *Guard the Chaos* write about the need for the transition stage of 'chaos' in the grieving process:

> for a new identity and a future that incorporates the past the grieving process must be entered into in all its stages. Most significantly there can be no leap from [orientation] to [reorientation], from the moment of loss to the new future.[12]

In former times, days were given for this process to happen and people were protected from the demands of ordinary life to give them time to grieve, which is the process of deintegration. Now the requirement that life go on as normal (when it clearly is not) means that when the self cannot stay integrated the process is pathologized, with the result that the person is now sick for a longer time and the self has huge problems in reintegrating, which it would do quite naturally, given time and support.

Psychologists like Thomas Fordham[13] and Joseph Redfearn[14] have challenged the notions of unity that underpinned Jung's concept of integration, while Assagioli[15] develops his notion of subpersonalities from Jung's notions of archetypes. The work of the philosopher Gillian Rose which includes the notion of working in what she calls 'the broken middle' has within it the necessity of living with the contradictions:

> her work seeks to retrieve the experience of contradiction as the substance of life lived in the rational and the actual. In the middle of imposed and negated identities and truths, in the uncertainty about who we are and what we should do, Gillian commends that we comprehend the brokenness of the middle as the education of our natural and philosophical consciousness. She commends us to work with these contradictions, with the roaring and the roasting of the broken middle, and to know that it is 'I'.[16]

To summarise this in a schema of the progressive breakdown of flow:

[12]   *Ibid.*, p. 16.
[13]   Fordham (1986).
[14]   Redfearn (1992).
[15]   Assagioli (1992).
[16]   Tubbs (1998), p. 34.

Creativity in the self ...

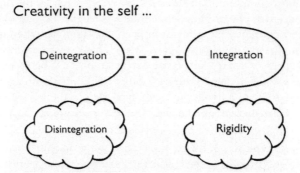

Mary Grey links this with religious ideas of the Underworld. In dealing with the story of Psyche and Eros she sees the problems Western society has with facing real life death and dying — the mythical underworld:

> Psyche returns from the Underworld — she has not conquered death in the way that Christians interpret the Resurrection of Jesus — but she has found a way of seeing in the dark, of living with death and destruction.[17]

Here it is important to remember the differentiation between deintegration and disintegration. Although the concept of fluidity in the self allows for this process of deintegration it does not mean the identity does not and cannot exist. The process of disintegration that many women experience imposed by external violence within a patriarchal culture can be differentiated from a process of deintegration lived through in an atmosphere of relationality and seen to be connected with the processes of deintegration that characterize the cycles of the earth.

To summarise, the pursuit of order over chaos, light over darkness, integration over deintegration has led to models of self and society that are repressive and authoritarian. The valuing of diversity and deintegration as a necessary part of self and society will enable organic change and cultural and personal creativity.

2 Public/Private

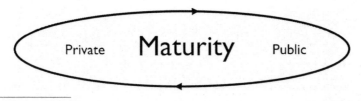

---

[17] Grey (1995).

The construction of this domain has had important significance for women's position in Western society. The binary of male=public, female=private (domestic) has been much discussed. The notion of privacy, particularly in relation to the press, is part of the daily discussion in the press itself. The British Royal Family struggle with it, particularly in relation to their children. We are in the West now in a situation where certain people — so-called celebrities — are always public (even in the most intimate parts of their lives), while certain people are completely private — the homeless, the differently abled, the criminals, to name but a few. Those 'in the public' eye will be valued by means of high salaries; the others may receive nothing. These latter have a hard time getting their experience known at all. They cannot express their thoughts, feelings or indeed the details of the difficulties of their lives. Repressive regimes have regularly used methods of surveillance, which make the private public. 'Bugging' was a regular part of Soviet Society. A society that loves publicity inserts cameras onto the most intimate parts of people's lives as the growth of surveillance cameras shows. The nightmare of the Big Brother of Huxley's *Brave New World* is becoming a reality and also a well-loved television show that people offer to be part of. With a high premium on the public, people seek publicity, without realizing how imbalanced their lives will become when they have it. So we have a society where in some situations like therapeutic ones, confidentiality is absolute. In situations that are abusive or inappropriately coercive in some way, this becomes twisted into an enforced secrecy and people are deliberately silenced. Free expression is often prized as an important value system of the West, but out of balance with an appropriate privacy, it leads to enforced or indecent exposure. To summarise this in a schema of the progressive breakdown of flow:

Maturity in society ...

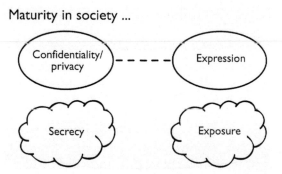

Within the self there are introvert and extrovert tendencies, even if one may appear dominant. A knowledge of when it is appropriate to speak and when be silent is a sign of maturity. The introvert who fails to keep in touch with his/her extrovert self becomes phobic; while the continual extrovert can lose all touch with more hidden part of his/her self. The slogan of feminists in the 1970s was that personal is political and this became a powerful rallying cry lining the private struggles of many women with wider socio-political issues. To summarise this in a schema of the progressive breakdown of flow:

Maturity in the self ...

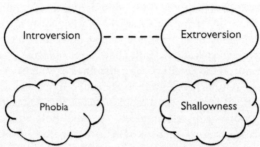

To summarise, Western culture prizes the public highly and devalues the private, which often has no economic value at all. While publicity is sought by many, there is clear line between those whose lives are almost totally public and those which are completely hidden. To enter the public arena in some artistic way is seen as beneficial and the movement of democratizing creativity has opened up the public space to more people. Debates about the position of subjugated knowers in both spheres have ranged from those who are happy to enter the public sphere as constructed by patriarchy and those who look for a radical redefinition of both spheres. The bringing of the two polarities into relationship could lead to a greater balance in people's lives (which is related to the work/life balance) and a greater variety of patterns of public leadership in the wider society.

3 Rational/intuitive

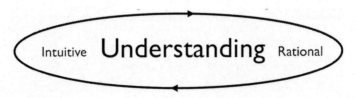

Post Enlightenment western culture has valued reason and devalued intuition. The Enlightenment project based on 'I think therefore I am' saw the answer to successful human society as the dominance of reason over human beings' unruly passions and imaginings. The intuitive aspects of the Church were suppressed in favour of theological codifications. Academe was suspicious of anything that smacked of the emotional or spiritual. Human beings were seen as being dispassionately objective about objects, situations and even other people. Disconnection between the observer and the observed was pursued and a culture of alienation and violence (in such methods as anatomical dissection) was developed. The emerging discipline of science and the older discipline of mathematics joined hands to develop methodologies that supposedly produced objective truths. These involved generalizing from large groups of objects and people by statistical methodologies (which often took little account of the fact that the groups chosen were very similar in ethnicity, gender or social class). This cult of 'objectivity' played a crucial part in tyrannizing groups of all kinds with notions of normality. The story, the oldest form of passing wisdom from one generation to the next, was devalued in favour of the lecture and in methodological terms stories were reduced to the 'merely anecdotal'. This culture passed from the academy into political judgements which are now supported by processes of number crunching, whether it is the controlled experimentation of the double blind control trials of the drug companies or the computed statistics of standardized attainment test scores that are use to produce leagues tables of schools. While there is a danger in seeing a single truth in a single story there is a cold intellectualism in seeing differences reduced to numbers and the natural world and human beings alike reduced to the status of alienated subjects. Emotional responses unchecked by reason leads to violence but cold principles have also destroyed vast tracts of the earth surface and peoples.

In this complexity, the role of the visionary experience is being re-evaluated and rediscovered. There is increasing interest in medieval visionaries like Julian of Norwich, Margery Kemp and Hildegard of Bingen. As one who regularly presents these women dramatically, I find it interesting that following many performances women, in particular will seek validation of their own visionary experiences. They will start tentatively with statements like 'I have never shared this with anyone before but ...' The telling of the stories of women of the past as a way of validating the experience of contemporary women has helped to redress the oppression of the intuitive response by the tools of the Enlightenment objectivity project. From an encounter with the spiritualities of the so-called New Age[18] comes a rediscovery of a working spirituality of angels, redressing their status as dusty relics of a bygone irrational age.

---

[18] For a summary of this phenomenon see Boyce-Tillman (2000), pp. 155–64.

So the challenging of the rational leads us inevitably to the mystical. The work of feminist theologians has been to redress the necessary balance in a methodology that validates lived experience as a source of truth. To summarise this in a schema of the progressive breakdown of flow:

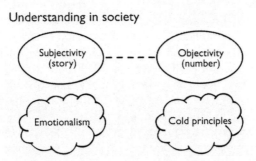

Understanding in society

Subjectivity (story) — — — Objectivity (number)

Emotionalism   Cold principles

Turning to the self, the visionary world, with its stress on imagination, is now repressed by years of persecutory educational enculturation. This world therefore appears in the unhelpful fantasizing and grotesque imagery that characterize both mental illness and the world of film and video alike. Alienated from their own imaginative potential, Westerners hand their imaginative capacities over to the big multinationals who manipulate it through entertainment and the techniques of advertising. Recently I saw a US game that is popular with children. It is an elaborate game with many roles within which the playing cards that feature monsters, dragons and erstwhile deities are asked to function in series of attacks and defenses. As the child related the rules to me, I wondered why he was not using the same material to create his own stories rather than relying on the rules of a large corporation that somewhere, I have no doubt, was making a great deal of money out of controlling what was once a free flowing human activity. To summarise this in a schema of the progressive breakdown of flow:

Understanding in the self ...

Visionary — — — Scientific

Lack of perspective   Cruelty

To summarise, Western rationalism has come to be at odds with intuition and its associated expressive and spiritual elements. The world of academe has passed to politics its support of so-called objective methods of evaluation. This has resulted in an imbalance in many areas and excessive alienation in the field of validated Western knowledge.

4 *Embodied/disembodied*

Western society, constructed as it was around the elements of (Greek-influenced) Christianity, had deep in its conceptualizing a notion of the body/soul split. The Enlightenment added a third element that could be split off — the mind or rational intellect. Few other human societies have achieved such an effective split between these elements. The consequences for life in western cultures have been considerable and these have been inflicted on cultures to whom they are absolutely foreign. Manual labour is now split from white and blue collar labour. One uses the bodies of people as if they have no minds, the other the mind as if it has no body. The ubiquity of computer technology requires most people to use a mind with minimal movements of the body which now has to be exercised separately almost with no mind. Work is for the mind and leisure for the body, with religion for the spirit if your cosmology includes one. The split leads to a division of power in society between those who exercise control through their decisions and those who implement the decisions through their actions, a division that has filled the corridors of academe for some time — only a limited number of radical thinkers actually use their bodies for social action. Politicians dream up political slogans but it is seldom they who put them into action. It is not politicians who fight in the world's battles or keep law and order on the city streets. The managers seldom 'get their hands dirty' on the shop floor. As Western rationalism still prevails, manual work, or action as I prefer to call it, is conceived of as having less value and the underprivileged of our society give their bodies to do essential practical tasks like cleaning, sewing and cooking, often in the power of someone else's mind. People are now prepared to sell the bodies for all kinds of purposes — whether it is its parts for medicines, or its sexual functioning in prostitution. Sometimes this is plain exploitation, sometimes it is pornography, and sometimes it is both. All of these can happen, when a person is conceived of as having a body but no mind or soul. The power divisions are ubiquitous: there are disembodied souls in the religious circles preaching a disembodied life; there are disembodied minds in political circles conceiving of principles and ideas with no basis in the lived bodily experience of those for whom they are conceiving them; there are disembodied minds also in academic circles generating principles which bear no relationship to practice, indeed which can be used as justifications for doing nothing because of the complexity of the problems involved.

The split of body from mind and spirit can be laid firmly at the door of the Church in such figures as Augustine, drawing on the Greek philosophies of figures like Plato. The excesses of asceticism are to be found in the stories of many medieval mystics and are now counterbalanced by the pornographic concern with the body (now minus mind and spirit) of the secular world. However, the development of Protestantism undoubtedly helped to widen the split further. As the sensuousness of Roman Catholicism with its visual images, its sweet smelling incense, its array of bodily gestures and varieties of music was systematically dismantled, the experience of God came to be located purely in the mind and spirit, and worship became denuded of symbols that involved the rest of the body.

Because the category of the transcendent in religion has become so enmeshed with the stress on disembodiment, the dilemma of whether of not to retain notions of transcendence at all runs through the feminist theological literature. Some would abandon notions of transcendence completely while others rework it in a variety of imaginative ways.

Jantzen retains a notion of transcendence to prevent 'reduction' of an embodied God into mere physiology[19]. Beatrice Bruteau's solution is one of 'cosmic incarnational mysticism' which restores to this polarity the fluid interchange of which it is capable:

> The ecstasy of the Theotokos[20] is transcendent of nature, but it is also continuous with nature, is itself perfectly natural. I would prefer to enlarge the concept of 'nature' until it includes all these transcendent acts of human consciousness, but without denying their transcendence or attempting to reduce them to notions of matter. Transcendence, emergence, and integration of the components are the very pattern of cosmic movement ... The incarnational model relieves us of the classical 'problem of evil' by not using its assumptions. Those assumptions are that God is both outside the creation and yet capable of acting as an independent agent inside creation,

---

[19] Jantzen (1998), p. 127.
[20] 'Mother of God' — the main appellation of the Virgin Mary in Orthodox Christianity.

changing some aspects of the creation without changing others. The incarnational model says that God is incarnate as the creation, thus acts as the whole of the creation in terms of lawful operations of that creation, and thus cannot (any more than any other agency) change some problems without upsetting others.[21]

The return of these subjugated knowers to a mystical view of God marks a stripping away of the male discourse that has overlaid the ample evidence in the Scriptures and later traditions that the mystery of God is beyond all comprehension.[22] Many theologians would see the sacred as being only within process and relinquish ideas of a stable metaphysical reality. This results in the move to verbs as expression of the Divine rather than nouns as we have seen above. Ruth Mantin refers to an interview with a woman who preferred to use the word 'Goddessing' as a more useful expression to her than 'Goddess'.[23]

The problems of 'keeping body and soul together' run through the literature and generate a variety of responses from theologians, theologians and philosophers. They balance the triumphalism of a patriarchal male God with a concentration on immanence and embodiment.

To summarise this in a schema of the progressive breakdown of flow:

Wisdom in society ...

Action ----- Belief/ understanding

Pornography

Dogma/ intellectualism

Within the self, the life of our society is reflected. Our lives are fragmented into activities which use either body, mind and spirit. It is perhaps a romantic idealization that says that an agrarian society saw people working in the fields at tasks that required all three in tune with one another. The concern for the body split off from mind or spirit characterizes such illnesses as anorexia and bulimia. These are but acute forms of a basic societal sickness. They have always been a part of women's experience of patriarchy. In the Middle Ages it was the ascetic excesses of the saints who in pursuit of a separated spirit abused their bodies. Today it is under pressure from the contemporary high priests

[21] Bruteau (1997), p. 176-7.
[22] Johnson (1993), p. 33.
[23] Mantin (2002).

of advertising or in response to experiences that have led them to regard their bodies as evil or polluted. Young women will have watched hundreds of millions of advertisements setting out ideal bodies by the time they reach puberty. None of them will have been presented with anything other than an idealized norm that apparently has no connection with mind or spirit. Young men will have been similarly enculturated to expect an obviously bigger and stronger body than a woman. Differences in sexuality will seldom be represented in the broader public sphere or increasingly groups will construct themselves around different images of the body of the one that it is appropriate to love. A narcissistic love of the body per se rather than for what it can do or express has been fuelled by an industry anxious to sell it expensive aids for making it more 'attractive'. Even sexual function has now been normalized. Following the discovery of Viagra for men, a corresponding medication has been invented for 'female sexual dysfunction'. This is based on norms established by heterosexual male sexuality. The woman who refuses to have genital relations with her husband as often as he would like is now dysfunctional; she is pathologised. Once upon a time there would have been a convent for her retreat to, where a vow of celibacy might protect her from unwanted advances. With celibacy discredited, largely because of its unnecessary alliance with the priesthood, there is now nowhere for her to retreat to except a guilty haze of pathological dysfunction. So does the dominant culture deal with its subjugated knowers. More people than the dominant culture would have us believe take temporary vows of celibacy after broken relationships (although they may not be expressed in those terms). One of the reasons I continue to teach the lives and writing of medieval mystic women is that I think it is a place that needs to be on the map — a place which the modern media will not publicize. There is no notion of variety in people's sexual needs and appetites either between different people of within the lives of the same person. To summarise this in a schema of the progressive breakdown of flow:

Wisdom in the self ...

The place of the concept of soul in our society is interesting. As church going declines a wider search for the spiritual fills the shelves of book-stores and health food shops. Much of it is related to bodily conditions, in a way that the church has lost touch with. Fasting was deleted in all but ritual form from Lenten practices only to find its way into detox programmes in the New Age dream. While incense was being down-played because of its sensuality people now burn it in their homes, while little bells tinkle at the entrance to houses to create good feng shui — bells that once would have signalled high points of a religious ritual. Our age is in search of a soul and there is a large industry seeking to provide it. It remains to be seen whether the gurus of the new spiritualities can effect a merger of body, mind and spirit better than the ecclesiastical authorities of the Christian church. Within Western society as it is it is still a basic human challenge to keep body and soul together.

To summarize, to keep body and soul together requires a relationship between them that roots the transcendent in the ordinary and everyday.

5 *Individual/Community*

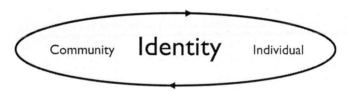

The relationship between the individual and the community has been expressed in different ways in different cultures. If the role of the community is paramount there will be often be a certain conservatism in the society maintained by the elders and a less clearly defined view of the individual self.[24] In societies with individualism prominent there will be a greater stress on freedom and innovation.

The legacy in the UK of the Thatcher years is one of fragmentation and an excessive emphasis on the individual. But it is a process that started with the Enlightenment and its rediscovery of the epic of the heroic journey. For many the heroic quest was never a possibility and women, for example, felt torn apart by a double standard — a society that prized the individualism that characterized the heroic search and their own sense of a deep need for community and stability for the sake of their children or their family.

Many people are seeing the possibilities of including connection with environment as a therapeutic tool for people trapped in the splitting of

---

[24] Floyd (1993).

this polarity. In the establishment of humanism at the Enlightenment, human beings became alienated from the natural world — or indeed superior to it. This resulted at best in a patronizing stewardship and at worst domination and outright rape. This was also linked with establishment of the individual self as distinct from other selves.

In a society where the polarity is splitting apart, responsibility will start to lose contact with personal freedom. It may be that certain groups of people will be scapegoated for the ills of society or certain people will take on the roles of being responsible. It is often women in the West who suffer from a sense of over-responsibility.[25]

When the polarity is finally split community can only be maintained by the coercive forces of law enforcement, and the state and its authorities will be regarded as oppressive of personal liberty. This will correspondingly be maintained by personal rights litigation. To summarise this in a schema of the progressive breakdown of flow:

Identity in society ...

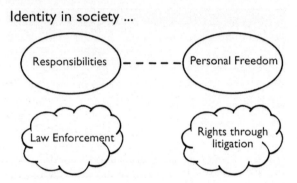

The individual experiences the pressures of the dominant culture in different ways. The dominant knowers in Western society will see personal liberty as of supreme importance and abrogate responsibility or see it as not part of their role. The subjugated knowers will be burdened by a sense of over responsibility and its contiguous guilt, often feeling trapped in situations that they feel they are unable to escape. When the split of the polarity finally occurs, the dominant knower will become isolated and aggressive, sensing a complete loss of any connection with the wider community which deep down s/he knows in necessary for his/her well-being. We see this phenomenon in young people turning weapons on their peers in school. The subjugated knower will lose any sense of boundaries and be overwhelmed by the burden of guilt and

---

[25] 'The challenge for men is this: Stop blaming: take responsibility for our life choices; and, in an ongoing way, hold ourselves accountable for our power' Ellison (1996), p. 113.

anxiety. The result may be pathologized anxiety, leading to self harm. To summarise this in a schema of the progressive breakdown of flow:

Identity in the self ...

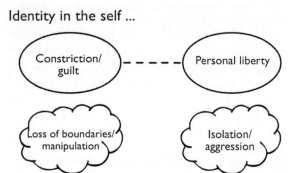

To stay in a job for more then five years is now regarded as inadvisable. Contracts are short. Yet the contract to bringing up children or care for a family is not short and it involves a great deal of repetition. There is often no sense of progress. When the washing up is finished there is another lot of plates being made dirty, when one garment is mended another is growing a hole, when one dispute is settled another one is brewing. Community building is not a heroic journey and it is the story of women in many cultures and of the poor in most cultures; it is often devalued and receives no returns in financial terms. The process of establishing and maintaining connections between people is a repetitive, circular journey. The loss of a sense of community with the natural world has resulted in the environmental tragedy that we are now facing. The earth and all non-human life has systematically been raped and pillaged in the interests of an aggressive individualism completely disconnected from its adjoined polarity of community.

To summarize, the bringing together of the individualism of the heroic journey with the community building enterprises of contemporary society will enable both individuals and communities themselves establish a true identity that is not dependent on a marginalized Other for its maintenance.

6 Product/process

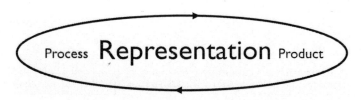

The sphere of work will tend to reflect the values of the dominant culture. Capitalism deals only in products. Those who are too old, too young, too ill or not skilled enough to produce products have no value and often little means to support themselves. This value system validates de-humanizing forms of production and the rape of the earth for the materials of production. Factory farming and sweat shop labouring are two examples of the phenomenon. People know that at work they are required to turn out particular sorts of things whether it is the cars of industry or the academic papers and books of academe. Non-production means redundancy. In many areas, this means a literate process carried on by means of written words in the form of email communication and reports and documentation of various kinds. The workplace is increasingly literate and oracy[26] can find no place in it.

In leisure we rebalance the demands of the work environment. Leisure activities are often much more concerned with process. The culture becomes more orate with conversation, whether over meals or in the pub, playing a significant part. So increasingly sedentary work is balanced by the gym and the running track. We don't do in leisure what we do at work. But leisure time is increasingly eroded for those at work. So we have one group of people for whom work consumes their entire life and one group who have no work at all. So the split opens up between the unemployed and the workaholics.

The division, above, into public and private is linked to the process/product split. The home (the more private sphere) can be seen as a process space where relationships are continually negotiated usually through the spoken word. The devaluing of process has done women no favours. The cyclical nature of a women's experience rooted in her own bodily processes means that she fits uneasily into the male constructs of the world of product based work. The western working structures with their patterns of reward and promotion have meant that women have been forced into bearing children later and later in their lives when the processes are much more difficult. Men too can pay little part in the process of bringing up their children because of the paramount demands of production. The advent of women in the workplace required a radical rethinking of the work structures of Western society which enabled the production based workplace to be balanced effectively with the process based domestic sphere. It would lead to

---

[26] The words literacy and oracy refer to processes of communication The terms literate and orate refer to societies which use either the written word or the spoken word as the prime means of communication.

- More flexible patterns of working
- Freeing up educational processes to enable child rearing to take place younger
- The re-establishment of networks of people such as the extended family and kinship groups to enable child rearing to be a communal activity
- The valuing of part-time working in career advancement
- The reduction of ageism in employment practices

Such a process would enable men and women to both 'be' and 'do' effectively.

Western society shows a valuing of literacy over oracy. In orate traditions the process of basing work on the product, and the itemization of time to defeat idleness, cannot happen — for the means of preservation is simply not available. There is an ephemerality about oral traditions within literate cultures; there is less desire to retain material from particular situations which may be inappropriate for different contexts. People who prefer oracy to literacy in our culture (and, in general, women are more skilled in verbal discourse than men) reflect this capacity to re-member material in ways appropriate to any given situation in which they find themselves.

Characteristics of orate cultures are:

- The absence of a definitive from of any story
- A fluidity in formal structures which are free flowing rather than linear and analytical
- Increased subjectivity
- Transmission by face-face contact not a book
- Open religion rather than a fixed revelation contained in a book.

To summarise this in a schema of the progressive breakdown of flow:

Representation in society ...

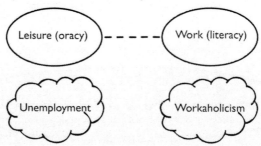

For the self there is ideally a balance between process and product. When being and doing become too split in a life or one gets left out altogether, then the system goes out of balance. Christopher Robin returning from his first day at school is asked by Pooh bear what it was like. His reply is that he thinks he will never be allowed to do nothing ever again. School is an initiation into the doing, the product based world of work. The end of it is burn-out, when the self finally rebels against the constant demand to produce and what's more to produce perfection. The idle person has regularly been condemned in the west, which has similarly condemned cultures with a greater degree of 'beingness' in their value systems as wasters and idlers.

To summarise this in a schema of the progressive breakdown of flow:

Representation in the self ...

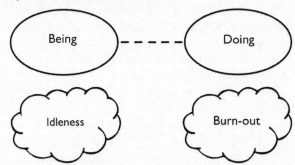

To summarise, when process and product are reconnected there is a balance in society of being and doing. Oracy and literacy will be equally valued so that there will be a concern for contextual considerations and a desire for appropriateness for context. Some marks of the process will be discernible in the product with a redefining of excellence that includes flaws. In theological terms this represents a balancing of humanity and divinity.

Limitations of space prevent discussion of the remaining polarities, which are merely stated below and discussed elsewhere.

7 Excitement/relaxation

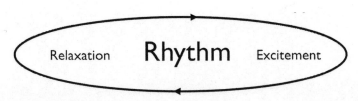

This is epitomized by

- The constant flickering images of the television set
- The constantly changing sounds of the walkman or the ubiquitously piped music which is often fast and loud
- The need to keep up with the latest fashion which changes rapidly enough to ensure that people will spend a great deal to keep up with the fast moving band wagon

The polarity breaks down as

Rhythm in society ...

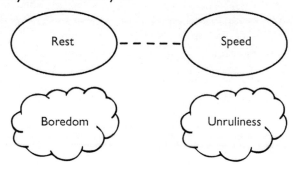

Rhythm in the self ...

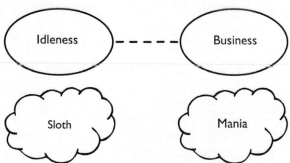

8 Challenge/nurture

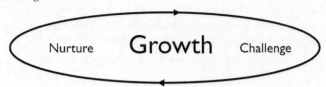

The capitalistic market is based on a philosophy that competition necessarily produces the best solution. Unfortunately separated from nurture it produces destruction as all sense of connectedness is systematically destroyed. Care separated from competition produces atrophy, and the separation of those who need care from those who can withstand competition has left many of the elders of our society atrophying in old people's homes without the challenges posed by association with other generations or the wider society.

Growth in society ...

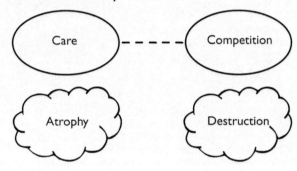

Growth in the self ...

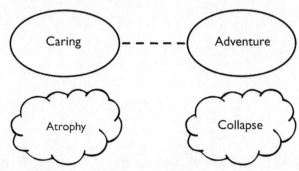

Overall Summary

This chapter has set out a model of how certain ways of knowing have become subjugated by the power structures of Western society. It has looked at a series of polarities which need to be continually flowing and in interrelationship and has examined the consequences of their splitting. It has called for the envisioning of new ways of restoring right relationship between the ways of knowing and the people who have embraced, or been forced into, particular patterns of knowing. It sees the need for a third point defining the connection of the two polarities. Hildegard expresses this well in her antiphon to Wisdom which she sees as having three wings:

> O the power of Wisdom:
> You, in circling, encircle all things,
> You are embracing everything in a way that brings life into being;
> For you have three wings.
> One of them reaches highest heaven
> And another is sweating in earth
> And the third is flying everywhere.
> Therefore it is right to give you praise,
> O Sophia wisdom.[27]

[27] Boyce-Tillman (1996).

John Holt

Creativity as the Immune System of the Mind and the Source of the Mythic

Self Realisation Through Symbolic Language

There is no sharp dividing line between self-repair and self realisation. All creative activity is a kind of do-it-yourself therapy, an attempt to come to terms with traumatising challenges. (Arthur Koestler)[1]

There has always been a human concern with the 'fractured self', right from the beginning of human consciousness. Questions fermented in the psyche: what constituted the fissure within and between minds and bodies, what is its causation and how do we repair the damage? And further, how do we reconcile the boundaries within us all, our environment and within the cosmos, between other beings and the spirits, and when things go wrong how do we rectify the imbalance, the illness? These dilemmas were addressed, until the development of the modern world, as spiritual problems with a divine process for healing. To be disconnected and dislocated from the notion of self in the deepest sense of the word is to be deprived of our potential as human beings and to be deluded into perceiving the world as a distorted and hostile place.

Self-realisation or self-integration can be said to be a tendency, a movement towards a greater understanding and acknowledgement of self in relation to the world. Self-realisation in this context does not imply total understanding, but more an improvement in the awareness of self and a reconstruction of awareness of one's relationship with the

---

[1]   Koestler (1975), p. 177.

environmental, personal milieu and history of the individual which charts the route of one's life journey. This was done using the templates of mythology as a guide. The journey is, in this very process, enhanced by self initiated and archetypal use of images and symbols which acts as a form of mapping in relation to mnemonic and re-collective experiences, and which also includes a future potential.

Ben Okri the African poet and writer has written widely on the purpose of stories to self and community. He values stories as powerful vehicles for understanding.

> When we have made an experience or a chaos into a story we have transformed it, made sense of it, transmuted experience, domesticated the chaos.[2]

We can begin by looking to the value of story telling, myth making, and art making towards the codification, the symbolisation of human experience. The need of human beings to understand their life experiences and to pass on their knowledge to others is, I suggest, a primary facility in many people and has been throughout time. Initially people who sought knowledge from the inner realms would be seen as shaman, medicine men and women, visionaries or mystics of some kind. Although this capacity is often seen as inherent and intense in some individuals, I think that the capacity for inner knowledge is a potential within every being. Also, the capacity to construct a complex language to articulate these experiences is, I believe, an inherent propensity in all people. Though language usages differ in form, quality and complexity of vocabulary, the fact remains that the mind wants and needs to formulate language even when it is not a major factor in the individual's own culture (in South America there have been examples of people with hearing difficulties formulating their own signing language when the social structures of their country broke down and there was no taught signing due to the social chaos). This process can be likened to a map making practice, only the human creative process of orientation is in constructing maps of life experiences, charting the journeys, the paths, the struggles, the disorientations and challenges of our pathways. Such processes were always valued and it was seen that those who experienced life deeply, often with pain and trauma, were those who had more to chart in their maps of experience.

Reconnecting to the Cosmos

One of the peculiar developments in our Western world is that we are losing our sense of the divine side of life, of the power of imagination, myth, dream and vision. The particular structure of modern consciousness, centred in a

[2]   Okri (1996).

rationalising, abstracting and controlling ego, determines the world in which we live and how we perceive and understand it; without the magical sense of perception, we do not live in a magical world. We no longer have the ability to shift mind sets and thus perceive other realities − to move between worlds, as ancient shamans did. Ritual signifies that something more is going on than meets the eye − something sacred.[3]

According to Suzi Gablik (quoted above) and philosophers such as Fritjof Capra (quoted below) we have lost our connection to the cosmos. Capra proposes the world view of systems theory in which the inter-meshing systems that abound in nature make every organism − animal, plant, microorganism, or human being − an integrated whole.

Being ecologically literate means understanding the basic patterns and processes by which nature sustains life and using these core concepts of ecology to create sustainable human communities, in particular, learning communities.[4]

Applying this ecological knowledge requires systems thinking, or thinking in terms of relationships, connectedness, and context. Eco-logical literacy means seeing the world as an interconnected whole. Using systems theory, we see that all living systems share a set of common properties and principles of organization. Thus we dis-cover similarities between phenomena at different levels of scale − the individual child, the classroom, the school, the district, and the surrounding human communities and ecosystems. With its intellec-tual grounding in systems thinking, ecoliteracy offers a powerful framework for a systemic approach to school reform.[5]

Throughout the living world, we find systems nesting within other systems. As Capra observes, living systems also include communities of organisms. These may be social systems − a family, a school, a vil-lage − or ecosystems or the system of an individual's psyche. Capra contends that as well as healthy systems, there are sick systems which produce an ongoing imbalance and fracture within their structures. Gablik also sees contemporary society as a force against the intuitive and sacred experiences of life and therefore a 'sick system' by definition.

In his significant and influential book *The Reenchantment of the World*, (which influenced Gablik's *Reenchantment of Art* quoted above) Dr Morris Berman, an academic, cultural historian and social critic, looks for a new world view that can give rise to a culture capable of relating gently and self-sustainingly to the earth and the people upon it.

This is not to treat mind, or consciousness, as an independent entity, cut off from material life: I hardly believe such is the case … The view of nature which predominated in the West down to the eve of the Scientific Revolution was that of an enchanted world. Rocks,

[3]   Gablik (1991).
[4]   http://www.ecoliteracy.org/pages/learningcommunities.html
[5]   Capra Quoted at: http://www.ecoliteracy.org/pages/fritjofcapra.html

trees, rivers, and clouds were seen as wondrous, alive, and human beings felt at home in this environment. The cosmos, in short was a place of *belonging*. A member of this cosmos was not an alienated observer of it but a direct participant in this drama. His personal destiny was bound up with its destiny, and this relationship gave meaning to his life. This type of consciousness — what I shall refer to in this book as 'participating consciousness' — involves a merger, or identification with one's surroundings, and bespeaks a psychic wholeness that has long since passed from the scene. Alchemy, as it turns out, was the last great coherent expression of participating consciousness in the West.[6]

And so it seems there was a disconnection from this 'participating consciousness' that Berman talks about. The question of this chapter is, how do we move towards this participating consciousness, this re-connecting, this self-realisation? When there is this abyss of disconnection there is often a desperate search for reconnection, as though it were a search for meaning.

It may not be of value, however, to fill whatever vacuum that seemingly exists in contemporary culture with either the exoticised notion of the 'primitive' or of the 'insane' as ultimate 'others' as was evidenced by the modernists and surrealists. Rather, as Berman, Gablik and Capra contend, there are models and paradigms past and present from which we can learn and apply to counter illness and imbalance of mind and body, much of which is exacerbated by the pressures of the contemporary world.

Creativity as the Immune System of the Mind.

As already stated, traditionally there were the healers, the shaman, medicine men and women, who maintained the balance through their knowledge of inner realms of experience, often sought by inducing psychotic states, in a search for the inner knowledge that requires continual renewal and new translations to remain vibrant and relevant. We no longer have the cultural context, it seems, for this kind of healing, as the rational reductionism of post Newtonian scientific view has thrown the 'baby out with the bathwater'. We are seemingly left alone, isolated in our unfulfilled lives, often getting to crisis point or simply breaking down before any help is given — and then it is to be treated with pharmaceutical and/or intrusive therapies. But so often the solutions are to be found from within ourselves. Just as the immune system of the physical body 'kicks in' when the body is threatened by bacteria, so it is with the mind when an acute state of imbalance is experienced: creativity comes forth to heal and to mend the 'broken hoop' of self.

[6]   Berman (1994).

Whenever illness is associated with loss of soul, the arts emerge
spontaneously as remedies, soul medicine. Pairing art and medicine
stimulates the creation of a discipline through which imagination
treats itself and re-cycles its vitality back into daily living.[7]

For this to happen, however, it is important that an empathetic environment is created in which this creative process of self-realisation can
be facilitated. This implies the creation of a space of trust and self own-
ership in which there is no betrayal of the individual's conviction in the
value of the process towards this end. This is specifically significant in
any case of the analytical use of such created images for clinical, diagnostic usage within a clinical environment.

'The Living Museum' at Creedmoor Psychiatric Center in Queens
New York demonstrates this. A unique exception to the medical model
of mental health, it was founded in 1984 by Bolek Greczynski, a Polish
avant-garde artist, and Dr. Janos Marton, a Hungarian artist and resi-
dent Creedmoor psychologist. They colonised Building 75, the former
kitchen and cafeteria, a decayed and decrepit outcome of twenty years
of downsizing policies at Creedmoor. Their concept was to create an art
museum with and for Creedmoor patients. This became a 40,000 square
foot art asylum, 'The Living Museum'. This is not 'art therapy' in the
sense of the emergent professionally trained 'art therapists': the work at
the Living Museum is never used for analysis or diagnosis, neither is it
directed or engineered in any way, it is just an artistic sanctuary in
which people labelled as mentally ill can re-define themselves as artists.

> Mental illness is one of the most horrifying conditions that a person
> can experience. But what I'm trying to do is to see the good and the
> positive aspects and to build on those. If you have delusions, for
> example, I see it as fertile ground for great art work.
>   The point is that art is not going to take away your mental illness.
> But it builds on the symptoms of mental illness. Great art really
> occurs in the domain where schizophrenic processes often occur.
>   In the regular hospital setting, that is something they'd want to
> get rid of, because that's what makes you crazy, what defines you as
> a mental patient, what gives you the diagnosis.
>   But I celebrate those aspects.
>   Most mentally ill people are over socialized by the time I get my
> hands on them. They have been told over and over again to do this,
> to do that, not to do this, not to do that. So in a way, pointing out
> rules doesn't help.
>   And second, which is that the rules are really self-imposed rules.
> If you own the place, when it's yours, if you set your own rules, you

---

[7]   Mcniff (1992), p. 1.

work along internalized rules. And it has worked fantastically. Nothing negative ever happened here.[8]

This positive, creative approach in many ways runs in the face of much that the mental health system stands for. The over emphasis on the pathological aspects of the individual to the exclusion of the goal of finding a personal orientation, of the need for an elevation of the spirit and self esteem, is largely ignored. But this should not be converted into an idealisation of self, a self-delusory train of thought: it is more a trust in the creative process to reveal the dualities of self in a natural and organic way.

> When therapy confines itself to treating 'sick people', its perspective is always of sickness. There is no opportunity for the pathology to have a productive interplay with other aspects of life, as we see in shamanic cultures.
>
> James Hillman helped me to see how my insistence that art therapy works with the 'healthy' and 'non-pathological' aspects of the person was a reaction to models of therapy that classify people according to pathologies. The classifying mind has used pathology to further its schemes. Rather then seeing pathology in a painting or dream as something sick and negative, we can embrace it as part of the soul's nature. This orientation does not take us away from the clinic, the hospital, and the prison but instead deepens our therapeutic identification with those places where the soul's suffering is most extreme ... This attitude assumes the wellness of the helper and sickness of the patient, and anyone who has spent time in a mental hospital as a patient or staff member knows that this is not the case ... When staff openly accept the pathology of their interactions with one another, then the system is ready for transformation.[9]

The construction of a lexicon of experientially based images and symbols can help to construct an ordered or re-structured understanding of self. Experiences of the individual in crisis are so often experienced as chaotic and random, seemingly having no reason, form or structure. The construction of a narrative, a story in whatever linguistic form begins to articulate the understanding. The self initiated and supported artistic process can begin to provide an order, a structure into which the individual can orientate and even plan future aspects of a personal journey devoid of the behaviours of the past but yet informed by the memories of the past.

This process is a form of mapping in which the practitioner is the visual or literary cartographer of her or his own psyche. This progression does not take place by means of a rational procedure, but more by

---

[8]   Dr. Janos Marton, Director of the 'Living Museum' Creedmoor Psychiatric Center, Queens New York. Quoted in the film *The Living Museum* by Jessica Yu (1999).

[9]   Mcniff (1992), p. 25.

an intuitive sense of the potential of such a practice. According to Arthur Koestler:

> To unlearn is more difficult than to learn; and it seems that the task of breaking up the rigid cognitive structures and re-assembling them into a new synthesis cannot, as a rule, be performed in the full daylight of the conscious, rational mind. It can only be done by reverting to those fluid, less committed and specialised forms of thinking which normally operate in the twilight of awareness.[10]

I have seen this work clearly in a group of artist-patients in a secure hospital in Britain where their lives, blighted by behaviours often driven by emotional and spiritual crisis with attributions in childhood, are slowly understood via the insights revealed by the creative process. As stated by Koestler, this is not an intellectual process, although the intellect plays a part as a reflective means; it is more an intuitive and creative release of images and symbolic shaping. Some of the artist-patients I have worked with have clearly indicated in a contracted exchange of artistic insight the significance of such a process to themselves. They talk of the channelling of energy and the positive cognitive value of their painting and drawing to their lives, incarcerated as they are in a secure hospital with all its connotations and its restrictions upon the human spirit and expression. These quotes have not been edited.

> 'I don't plan — nothing. When I work it just comes naturally and none of it is thought out. I couldn't do this work every day but I value it once a week. If I were to come more I wouldn't get the emotion because I would be emotionally drained. I would have to force my emotions out.'

> 'My influence was a mate in prison who had done over twenty years. He found inner peace from his art-work. He used to be very violent, but he found peace with art. Artists used to visit prison to see his work and he showed them some of mine. They saw things in my work I didn't know were there. Now I find that working here gives me inner peace.'

> 'Consider a desert, barren and empty, then rain comes and the water gives existence to the once lifeless plain, and, oh, what life can come from the sky above.
>
> Just as the rain gives life to the desert, we can give life to existence, and from existence the journey of meaning can be travelled; the journey we can measure, should we choose, through the record of internal dialogue via Art.
>
> The reflection of the joy, the sorrow, despair, anger, hate, fear, love, stillness, can be captured in a moment by the contact of brush on canvas.
>
> We can embark on a truly amazing voyage of self-discovery, being a love affair within.

---

[10]   Koestler (1975), p. 179.

We can look deeper into ourselves and see the pain, turmoil and fear, as well as the joy, love, ecstasy of being, our conflicting beliefs and mesmerising dreams of power beyond imagination.

The development of Man is to be measured in how he pushes forward, how he overcomes seemingly insurmountable obstacles, and defeats the darkness with a will of iron; how he steers his course through the icy winds of destiny, conquers his fears and tastes the sweet joy of victory by carving the "I" in his identity on top of the jagged mountain of his dreams.'

'All the emotions, love, peace, hate, war, death, the Devil, sin, the tattoos — it's deep.'[11]

The Creative Response to Suffering

As some of the coherent comments of the artist-patients above contend, this is, however, not always a comforting and peaceful process, as it can raise and recall painful and traumatising experiences. The process is often then shut off in a self regulatory way to the extent of what can be borne, what can be tolerated at the time. The return of the muse or the very reason for its initiation can be described as an act of faith in the creative process itself, driven by what Wassily Kandinsky defined as determined by an 'inner-necessity'. This faith in the creative process seems to be strong in many people suffering disorientation in their lives.

> A curvilinear relationship exists, between conflict, pain suffering and creativity. Some conflicts internally felt by the creator may be projected onto his own creation and lend it a personalised dimension of depth. Too much conflict, however, may stifle the creator. In a similar vein, straining to overcome pain, interference and obstacles fuels the fires of creativity, but unbearable pain and insurmountable obstacles are liable to extinguish them. Either we transform our burden into something creative or slump under it without authenticity, confounded by a petrifying routine ... Since creativity is the prime mode of communication it might well be the antidote to violence. (Shoham 2003)

I do not advocate the seeking of suffering as the means to achievement. What the mind actually seeks is *peace* and yet what it gets is so often suffering. The Buddhists would contend that all life is suffering and it is from the premise of this understanding that we learn not to attach to the pain of suffering, but to seek a pure and therefore peaceful mind. The antidote to violence, which is in itself born of anger, as cited by Shlomo Giora Shoham, is shown by my work in a Special Hospital in which patients, ridden with anger and violence towards themselves

---

[11] Quotes given by artist-patients as a consequence of a symposium on art and meaning in a secure hospital. (Art and Mind, 2001).

and others, often begin to understand the underlying nature of their inner hostility through their self managed creativity.

> I've got a more peaceful outlook on life. I used to draw really aggressive, nasty pictures, but now I like atmospheric pictures. I don't think I could sit down and draw an aggressive picture now, and it come out like it did eight years ago.
>
> I have been painting for about a year, but only twice a week, and then infrequently because of my health. Since I have started to paint I am aware of images all around me, on the wallpaper, and in the 'floaters' in my eyes.[12]

Artistic Creation as Ritual

The means of approach to self-realisation is one of small or large steps towards this state of awareness. It can be compared to the construction of a personal "rite of passage" and as such sometimes the removal of the barriers to revelation can be overwhelming or difficult to absorb all at one time. The notion of "ritual art" is one which is particularly evidenced in South Asia where the transcription of psychic processes teem out of the rich and vital iconography of the minds of the Vedic traditions into the incomparably visually rich visual languages of Hinduism and Buddhism. In this following definition of ritual art Ajit Mookerjee places the realisation of cosmic connectedness and ultimate self realisation not in a distant holy place but within every being:

> Ritual art is a means or way towards spiritual identity, towards a state in which we can realise our oneness with the universe. *This realisation is not something that descends from above; rather, it is an illumination to be discovered within.* The unity underlying the diversity of the world is to be discovered in our relationships with all life, manifest and non-manifest. Integration of the self is achieved in ritual worship which opens up contact with each and very aspect of our being. Ritual works on the assumption that nothing, however small and apparently insignificant, or vast and incomprehensible, is without significance to our destiny in this *jagat*, the ever-moving world.
>
> The arts have their origin in ritual. Whether to invoke and propitiate deities, exorcise negative forces, to celebrate rites of passage or mark turning points in the cycle of death and renewal, ritual creates a focus and compacts energy.[13]

Notions of personal mythologies work if we see myth as the maps of individual journeys which are so significant in pre-industrial and tribal or shamanic cultures. All art, it can be argued, is a form of mythology, a symbolic clarification of our experiences of the world, and of our relationships with the world. To correct a damaging mythological path, one

---

[12]   Patient in Special Hospital.
[13]   Mookergee (1998), p. 9.

which can lead in itself to a personal crisis and collapse, one has to open up to the process of creation itself to allow the self-regulatory process to work in a meaningful way:

> Much of the psychological suffering people experience is entangled in personal myths that are not attuned to their actual needs, potentials and circumstances. Attempting to follow a personal myth that is not in harmony with who you are or with in the world in which you live is painful, and a mythology that is unable to serve as a bridge to deeper meanings and greater inspiration than you can find in the outer world is often accompanied by a deep and nameless anxiety. As you develop greater awareness of your emerging mythology, you experience increased intimacy with your inner being. To 'know thyself' in this informed and substantial manner inoculates you against at least one strain of the generalised anxiety of the day by day, engaging greater support from your deeper self.
>
> Myths, in the sense that we are using the term, are *not* legends or falsehoods. They are rather, the models by which human beings code and organise their perceptions, feelings, thoughts and actions. Your personal mythology is rooted in the very ground of your being, and it is also a reflection of the mythology held by the culture in which you live. We all create myths based upon sources that are within us and sources that are external, and we live according to those myths. Psychologist Henry Murray thought of myths this way: myths serve to inspire, generate conviction, orientate action and unify a person or a group by creating the passionate participation of all functions of the personality (*individual* myth), or of all members of a society (*collective* myth).[14]

The arts as a means of a 'compacting of energy' is seen clearly in some of the mystical outpourings of people across diverse cultures and throughout histories. One such individual is the Spanish mystic poet John of the Cross who wrote poignant revelations of his experiences of the routes to aspects of inner self, albeit deeply steeped in Christian imagery and symbolism. His sense of trauma at what he experienced as a loss of God's presence (loss of orientation towards the light of understanding) of the process of self-reconstruction and sensitisation and his sense of movement, of passage through layers of consciousness, indicate an acute articulation of the creative experience at its most intense state. This poetry could be seen by contemporary psychiatric practice as pathological in nature, revealing delusory and psychotic tendencies, but to many it reveals an articulation of states of transcendence, of union.

> I entered in, I know not where,
> And I remained, though knowing naught,
> Transcending knowledge with my thought.

---

[14] Feinstein and Krippner (1988).

Of when I entered I know naught,
But when I saw that I was there
(though where it was I did not care)
Strange things I learned, with greatness fraught.
Yet what I heard I'll not declare.
But there I stayed, though knowing naught,
Transcending knowledge with my thought.

Of peace and piety interwound
This perfect science had been wrought,
Within the solitude profound
A straight and narrow path it taught,
Such secret wisdom there I found
That there I stammered, saying naught,
But topped all knowledge with my thought.

So borne aloft, so drunken-reeling,
So rapt was I, so swept away,
Within the scope of sense or feeling
My sense or feeling could not stay.
And in my soul I felt, revealing,
A sense that, though its sense was naught,
Transcended knowledge with my thought.

The man who truly there has come
Of his own self must shed the guise;
Of all he knew before the sum
Seems far beneath that wondrous prize:
And in this lore he grows so wise
That he remains, though knowing naught,
Transcending knowledge with his thought.[15]

As the patient so eloquently states in his observation 'Consider a
desert ...'[16], the sense of abandonment and vast voids of featurelessness
of St John's notion of 'shedding ... of all he knew before the sum' is
achieved in relation to our perception of self-realisation through the
creative, poetic processes we are advocating here. St. John does not
idealise this inner journey of an eventual 'transcendence of self (knowledge)', more he points to its 'strangeness' which left him 'stammering'.
His poetic dialogue is the linguistic manifestation of this experience of
movement towards a unification of self. In subsequent poems he
propounds the notion of 'I live without inhabiting Myself — in such a
wise that I Am dying that I do not die'.[17]

---

[15]   'Verses written after an ecstasy of high exaltation', John of the Cross (1979).
[16]   Above, p. 40.
[17]   'Coplas about the soul which suffers with impatience to see God', John of
       the Cross (1979), p. 35.

This poem indicates the possibility that one can release oneself from a previous state of being, of thought and perception and move into a different perspective liberated from the acknowledged prior or previous state. This could be defined as a 'rite of passage' in which, in this case, an induced state of heightened consciousness leads to a shift in the way one sees and engages with the world.

Subjugation by the Medical Model

The use of powerful anti-psychotic and anti-depressant drugs often inhibits and blocks the creative process in the individual. This factor so often causes distress and concern in the individual, a concern that does not seem to be of significance in the diagnostic and clinical process. I have had examples of patients complaining about the loss of their creativity when their medication is changed or increased. This caused distress to the patients who could no longer participate in the illumination of their creativity, and when I suggested that they mention this to their psychiatrist they said that they had done just that, but it was of no consequence to the clinician that they had lost their creativity. Inherent in all this is a political dimension in which those with vested interests in the dominant model fear a loss of power and control from any alternative, more creative and holistic model of healing being offered — particularly one in which the integration process is self directed. A holistic and creative approach to the care of those in crisis is not a majority occurrence in my experience, and where present it is seen as marginal and merely superficial to the clinical implications. There are, of course enlightened and radical programmes of care, but they are not in the mainstream or are short lived initiatives, often led by clients themselves without support from the dominant institutional structures.

As observed by the director of the 'Living Museum', Dr. Janos Marton, the creative, artistic work of those who are defined as mentally ill has significance for society as a whole through the often profound understanding that this work affords in giving a deep insight into the understanding of the psyche. In the sharing of 'one's own misfortunes' Indian academic Ananda Coomaraswamy says something significant is given to the world.

> Each race contributes something essential to the world's civilisation in the course of its own self expression and self-realisation. The character built up in solving ones own problems, in the experience of its own misfortunes, is itself a gift which each offers to the world.[18]

---

[18] Coomaraswamy (1985).

This exposure of the images of self revelation should not have subjective censorship applied to it. Censorship creates taboos and psychological 'no go areas' which in themselves are often the very cause of trauma, blockage and fracture in the first place. Misinformed and misdirected censorship is in itself a form of oppression of the psychic process and plays into the hands of reductionist and suppressive viewpoints, often with their own agendas and self righteous platforms of hatred and repression; or it panders to scientific views which reduce the symbol to pathology. Once we suppress freedom of expression and freedom to view the image we are lost in a dark world in which images cannot be brought to light, but are hidden in a vault to mutate and reappear in often physically violent ways.

As already stated this revelatory process of self-realisation has value not only for the creator, but also for the audience of such work. Renowned American art therapist Shaun McNiff writes: 'The best medicine I can offer to a troubled person is a sense of purpose, the feeling that what he is going through may contribute to the vitality of the community.'[19]

Touched, it seems by the 'humanity' and revelatory aspects of such imagery the public (community), seem surprised by the depth of effectiveness of such visual metaphors, initiating a response touching upon self-realisation and awareness in the viewers themselves. The following quotes are taken directly from the comments book of a public exhibition of over seventy paintings and drawings produced by artist-patients from a secure hospital in Britain exhibited in 2001 (Art and Mind). They have not been edited:

> 'Thank you for sharing your thoughts, feelings and images with us. I found this exhibition very powerful and moving. It touched my heart and mind. Congratulations!!!
> May this work continue to flourish and blossom and may you find your own inner peace through this wonderful process.'

> 'Some of this exhibition felt very personal & I appreciated being around to see them. The paintings had many effects — some inspired me some frightened me some gave me the desire for peace — art is a wonderful thing, thank you.'

> 'The pictures were a release. Thank you all for your expression. Images are so powerful. They echo like ripples on a pond in the mind of the observer.'

> 'A breath of fresh air. Thought provoking through the simplicity and subtlety of the artwork. Overall a very important exhibition giving a real insight into another way of seeing.'

---

[19]   McNiff (1992), p. 25.

The Recovery of Art and the Art of Recovery

Carl Jung talked of humanity's image making propensity and how this primary form of outer and inner experience has dominated human thought and created the phenomena of symbolism and mythology:

> The history of symbolism shows that everything can assume symbolic significance: natural objects (like stones, plants, animals, men, mountains and valleys, sun and moon, wind water and fire), or man made things (like houses, boats, or cars), or even abstract forms (like numbers, or the triangle, the square and the circle). In fact, the whole cosmos is a potential symbol.
>
> Man with his symbol-making propensity, unconsciously transforms objects or forms into symbols (thereby endowing them with great psychological importance).[20]

I would ague that this 'image making propensity' is inherent to humanity, but that the cultural and social expectation, the educational system and the very ethos of contemporary society conspires against the primary need to be creative. I have also spoken to people from other ethnic backgrounds who told me that parental suppression of creativity caused mental ill health within them. It is often only in a state of crisis that people begin to engage in the creativity that is dormant within; as though knowledge and the capacity for self-realisation was inherent and knowable through a process of revelation in the course of accessing the pathway to the process of creativity.

> Even at the start of the twentieth century, Prinzhorn was writing: 'There exists a primitive need in every human being to create which the development of civilisation has obscured'.
>
> 'It is not genius, contrary to what everyone says, which has ever prevented the production of art' echoed Debuffet in the preface to the first *Cahier de L'Art Brut* in 1964. 'Genius is the most fluid currency which exists, genius abounds and any newcomer has sufficient reserve to produce a work of art.'[21]

I have encountered a number of artist-patients in secure psychiatric hospitals who had never thought of themselves or the arts as significant until they were incarcerated in an institution. This development came about not as a result of clinical application but more as part of an optional educational programme. Once they had encountered the value of the creative process it seemed to assume enormous significance in their lives. This can certainly be said to be true for the artists of the 'Living Museum'.

> But then suddenly you think of yourself as a painter, as a poet. You can, realistically, too, because the Living Museum provides you

---

[20] Aniela Jaffé, in Jung (1980), Ch. 4, p. 255.
[21] Danchin (1989).

with a realistic framework to do that. You exhibit. People come to
see your work. Suddenly, the disadvantage of your existence, which
is time and boredom and waiting — that's the reality of mental illness, endless waiting, endless emptiness, endless void — suddenly
turns into an asset: 'Oh my God, I have time to paint!'[22]

It is difficult to measure the shift from disorientation to the sense of
orientation afforded by creativity. It can be validated orally and by
example but research is thin in relation to the transformation of status
and understanding that creativity can give to an individual.

Self realization through the creative process should, it seems, be seen
as a means to the 'domestication of a chaos' that Ben Okri propounds in
his notion of narrative and stories.

In an article in *The Guardian* on December 31st 2000 entitled 'Harsh
words can deform children's brains for life', Dr Martin Teicher, who led
a study team at McLean's Hospital in Massachsetts, said there was
evidence that emotional trauma in childhood could cause deformities
in critical parts of the brain, in turn causing depression, anxiety and
other conditions.

Teicher believed his research points towards a different type of treatment in which creative processes involving concentrated co-ordination
of both brain hemispheres could lead to a regeneration of some damaged parts of the brain, and could be a potential cure. This reminds me
of the Taoist notion of re-connecting with the essential energies in
nature as though we had forgotten to connect. The disconnection of
aspects of the brain through abuse and the need to protect the emerging
mind is as though the mind had 'forgot' its own creative potential. The
encounter with symbolic language in creative functioning allows the
mind to reconnect to the dormant aspects of the mind, hence reactivat-
ing and re-integrating the individual with the creative forces both
within and without the psyche.

Although this may seem to be a consideration of arts provision and
specifically the visual aspects, the whole intention of this chapter is
directed to the application of creative process to people who find themselves in crises, both emotional and spiritual, and this should have a
practical and activist implication. The scientific mind needs validation
of these theories before it commits its resources to the facilitation of creative sanctuaries advocated by such organisations as AIM (Artists in
Mind) which works in both hospital and community settings providing
sanctuaries, free from pathological and judgmental analysis in which
the map making we are advocating here can take place.

---

[22] Dr. Janos Marton, Director of the 'Living Museum' Creedmoor Psychiatric
      Center, Queens New York. Quote from the film *The Living Museum* by
      Jecssica Yu (1999).

Evidence collecting should include assessing the quality of artwork and being sensitive to people's development potential ... No systematic review of the social and clinical costs and benefits of participation in arts and mental health programmes appear to have been undertaken.[23]

Work produced in the process of self realization is not second rate or merely indulgent art. It can be of the highest standard and quality, in western art terms, and such collections of art as the Prinzhorn Collection in Heidelberg in Germany will testify to this. But as with all expressive skills there needs to be time to mature and develop a succinct and articulate language, often one which breaks the barriers and the rules. One young Muslim artist with a history of mental illness told me 'You have to break the rules for God!'. And so the artists who suffer disconnection make art, it seems, for a singular purpose; with, in the first instance, total absorption in the process of imaginative reconstruction. The notion of audience seems to be an afterthought, but one which takes on significance when validation is sought for the visions of their interior worlds.

Art does not lie down on the bed that is made for it;
it runs away as soon as one says its name;
it loves to be incognito.
Its best moments are when it forgets
what it is called.[24]

There is, however, in the contemporary world a need for categorization and codification connected to the notion of the search for meaning. As I have already mentioned, this led to what became the exoticisation of insanity. This genre, entitled 'Outsider Art' or 'Art Brut' took the marginal figure, ones who seemed to embrace encounters with other realities, and thrust them into the position of psychic pioneers. As I wrote in my article 'Anthropologists of the Mind':

'Anthropologists of the mind' have their own territories of colonisation and exotic analysis and just as anthropology went through its crisis of identity, so I feel that 'Outsider Art' should re-examine its premise in a more consultative and inclusive way with those upon whom their gaze falls.[25]

Creativity as the immune system of the mind and the source of the mythic is a premise that permeates my own work and the ethos of this chapter. I believe that the creative process in the form of the arts in its full range is the location of the true expression of life's diversity and of life journeys, indeed of philosophy itself. To make meaning of our lives and construct mythological maps of our journeys moves us to a deeper

---

[23]   Cowling (2004).
[24]   Jean Debuffet, in Hayward Gallery (1979).
[25]   Holt (2003).

understanding of self and the relationship between our minds/bodies and the environment. This process of re-connection with the cosmos is inherent and eternally human. I do not believe in the reduction of humanity to judgements of pathology, deviance and aberrations from what is considered 'normal'. A reductionist model is in no way useful in reconnecting or re-balancing minds which have been damaged in any way. People who share their own 'ways of knowing' through linguistic constructs reveal a consciousness which is heightened often in frightening ways by its struggle to comprehend, to find its own peace in other than conflicting and intrusive ways. From the platform of this work the viewer, the audience, reconstructs those maps of the psyche in ways appropriate to their own passage through life, using the signs and symbols in significant ways for themselves. And so this process heals at each stage, in the making and in the sharing.

I do not advocate this creative openness merely for those in crisis, although its value to people who have lost their way and become imbalanced to their own needs to fulfilment are inestimable. No, this is an entitlement for everyone. The knowledge that creativity is the immune system of mind, body and spirit is a resource for everyone and one that needs to be valued for the future health of all.

And from those who have survived the mental health system there is so often a story to tell, a message to give. One such survivor, Angela, thinks of her long incarceration in various psychiatric institutions as a process of spiritual transformation, although her period of deep experiences was defined as 'paranoid Schizophrenia' and not afforded the significance of a spiritual journey that she herself contended. Her creativity won through and she did survive after many years of drug and therapeutic intervention. She wrote with great courage a clarion cry not just for those defined as mentally ill but for us all:

> Those of us with a grander vision of human possibilities will refuse to sanction as reality this vision of a soulless universe that has cast a spell upon the collective imagination of humanity. We refuse to collaborate in this process: 'the disenchantment of the world and its transformation into a causal mechanism'. We will protest, we will fight — in the name of the Imagination, in the name of Love, in the name of God, in the name of madness.[26]

---

[26] Farber (1993), ch. 6, 'Angela's Story', p. 98.

Jennifer Elam

Soul's Sanctuary

Mystical Experience as a Way of Knowing

The Narrowing of the 'Normal'

The most beautiful and profound emotion we can experience is the sensation of the mystical. It is the source of all true science ... To know that what is impenetrable to us really exists − this knowledge, this feeling, is the center of true religion ... All true art, religion, and science come from the same place. (Albert Einstein)[1]

According to Albert Einstein, there is no conflict among true science, true religion, and true art. They come from the same source of creation energy: the mystical. Is that source available to us today? Continuing revelation would seem to be the common belief system between science and mysticism. How exciting are the possibilities when we realize that we can be the vessel for carrying the messages of creation forward to our generation! It may not be in the same celebrated and acknowledged way for which Einstein was known, but I believe that we *can* each bring our own little piece of creation to the world − if we are open to that possibility.

I sat in Quaker meeting one Sunday morning as Marjorie told us that her fourteen-year-old son, David, had made a great discovery related to computer technology. She was worried about how to manage the lawyers, the funders, and all the people that would enter their family's life because of this magnificent discovery. Later, I wondered how David happened upon this discovery. In my inquiry, I learned that David was home-schooled and had not been made to conform to 'the box' of learning to which most children must conform. As a result, it had not been

---

[1]   Quoted in Bucky and Weakland (1992), p. 86.

hard for him to think outside the box or to be open to larger possibilities. His 'discovery' seemed to be the result of the creation spirit working within a child that was open.

It is sometimes devalued, ignored, suppressed, or misunderstood. But mystical, creation energy is available to us today. Revelation continues.

Many people choose to live in small boxes, in small worlds, cutting themselves off from that energy, that knowing. Others keep their visionary experiences to themselves out of fear of being called 'crazy' or being formally labelled with a 'mental illness'. Yet some researchers estimate that those in Western cultures who have had what might be called a mystical experience is somewhere between 40 and 90 per cent.

In my own research on mystical experiences, I listened to about one hundred personal stories. From the third person who spoke with me to the ninety-third, I repeatedly heard: 'I have never told anyone this story about the mystical in my life. People would call me crazy.' From my research, I concluded that many people do not feel safe sharing their experience — and do not ask for the help they need to navigate the waters of the mystical. In addition, when those mystical experiences are suppressed, pathologies are created.

As our world narrows the definition of 'normal' and as Western cultures, in particular, assign more and more kinds of experiences to the realm of pathology, many people are shutting down their visionary experiences. They are cutting themselves off from rich ways of knowing. Society's squashing of the mystical often begins at an early age. Even in preschool, what is acceptable has narrowed. Buying and wearing the 'right' brands of clothing is important. Children are often ostracized if do not fit into the narrow band that commercial advertisers in the mass media have fostered as 'normal'. Cause and effect is difficult to prove, but when they grow up, many children who have encountered this kind of rejection seem to have trouble fitting into 'normal' worlds of relationships and work.

Children who do not fit into predefined patterns of what is 'normal' are not 'sick'. They are not 'wrong'. They are just different. Instead of fostering their unique gifts, schools create pathologies for them, and then nourisht and develop the pathologies. Instead of making a child's educational environment more healthy and tolerant, schools focus almost exclusively on making each child fit somehow into the institutions, however intolerant those institutions might be.

Mystical Knowing Through the Ages

All shall be well, and all shall be well, and all manner of things shall be well.
(Julian of Norwich)[2]

The place that I have found where art, religion and science can come together compatibly is in the valuing of mystical experience as a way of knowing. The greatest scientists — as well as mystics and artists — have opened themselves to knowing what they are given to see, hear, and feel even when others do not see, hear, or feel the same.

Many mystics through the ages have given us their 'knowings'. They knew because they heard, they saw, or they felt the presence of Spirit with them. Among the most celebrated women mystics — who knew what they knew through mystical experiences — are such people as Julian of Norwich, Teresa of Avila, and many others. Each lived an inner life that led to intimacy with Spirit. But once they knew what they knew, their lives moved outward. They had to share their knowings with the world, and because they shared, many throughout the world have benefited from what they learned through mystical ways of knowing.

Julian was the mystical and reclusive Abbess of Norwich in Norfolk, England during the fourteenth and early fifteenth centuries. While at the point of death, she experienced numerous visions and then made a startling, rapid recovery. Following her recovery, she concluded that her visions must have been delirious ravings. However, she then had another vision in which she was told that her sick-bed visions were genuine spiritual revelations and that she was to share them with others.

For decades following these visions, she continued to receive divinely inspired teachings which helped her to interpret the meaning of the initial visions. She spent her life writing about her visions and what she had been shown. Julian was filled with hope, teaching us that the main message of the Divine to us is that everything is going to be all right. She also is the foremost feminist mystic, relating to the Divine as feminine. Julian taught us that Christ is our Mother and that we are spiritually nourished at Christ's breasts.

Saint Teresa was born in Avila, Spain, in 1515. Twenty years later, she entered the Carmelite Monastery of the Incarnation at Avila, her father resigning himself to this development. Shortly after her profession, however, Teresa became seriously ill, failing to respond to medical treatment. On August 15, 1539 she fell into a coma so profound that she was thought to be dead. After four days, she revived, but she remained paralysed in her legs for another three years.

After her cure, which she attributed to St. Joseph, Teresa entered a period of aridity in her spiritual life. Even then, she did not give up

[2]   Julian of Norwich (1977), ch. 27, p. 124.

praying. Her trouble came from not understanding that the use of the imagination could be dispensed with and that her soul could give itself directly to contemplation. During this stage, which lasted eighteen years, she had transitory mystical experiences. She was held back by a strong desire to be appreciated by others, but this finally left her in an experience of conversion in the presence of an image of "the sorely wounded Christ." This conversion dislodged the egoism that had hindered her spiritual development. Thus, at the age of thirty nine, she began to enjoy a vivid experience of God's presence within her. Teresa spent much of her life working for reform of her faith community and wrote extensively on many subjects, especially prayer.

Contemporary Mystical Knowing

Painting is a form of playing and praying. Soul is being made visible. In painting, I am creating worlds of possibility. I am co-creating with Spirit; my hands are tools being used. Prayer leads me into this type of painting more than technique. I am allowing the electromagnetic energy that is our physical conduit for Spirit to flow through me and out my hands. I am letting go of control to the greatest extent possible. I may be meditating on my breath. I may be meditating on a dying friend. I may be meditating on prayers of gratitude or intercession. I may be asking for help. Whatever my meditation, I seek lightness of heart without invalidating my heavy heart. In all, my pure, raw, mystical energy bubbles forth. 'Remember that the space within,' she says as we paint, 'is as vast as the sky above.' (Jen Elam)[3]

I sat in the circle. We had done art together all day, meditating. After inviting the Spirit of Creation Energy to join us, we worked in silence. We painted, we wrote poetry, we made books, and we wrote the stories of our hearts. Our circle was a healing place. One person had breast cancer; one had chronic depression; one had heart disease; others had terminal illnesses of various sorts. We came together to embrace healing and joy.

When asked what brought joy to our hearts and made our spirits dance, most said children, nature, animals, or art. When asked to share the stories of what brings joy in ways that others could see, hear, taste, smell, or feel, the stories came alive with palpable, transmittable joy. Suddenly, one could feel a mystical presence in the room, delighting in the circle of healing. In that place of child-like wonder, the mystical emerges, creation emerges — and destruction, violence and war become impossible.

Mystical creative energy is not just something of the past. It is alive today. Many people continue to open themselves to those possibilities, and as a result, they begin to *know* that which is beyond daily life. They

---

[3]   Elam (2005).

begin to experience something of the numinous that is ineffable. And more often than not, this *knowing* invites something new to be created into the world.

Underlying all of the major religions of the world today, you will find, is a mystical tradition. Mystics continue to see, hear, and feel things which inspire them to begin something new. In addition to the major religions, there are native ways, pagan ways, and feminine ways of knowing. But all of these as well can be traced back to someone, somewhere, opening themselves to the mystical and hearing, seeing, feeling, tasting, smelling and *knowing* a new way of being.

By way of example, let me share some stories (abbreviated in places) that have been shared with me. As you will see, today's mystical manifestations of Spirit are many and varied. Yet each story is a reminder that even today, there are ways of knowing that go far beyond that which is typically defined as 'normal', ways of knowing that go far beyond the consumer-oriented routines of daily life.

Tom's Knowing: 'Beyond the head'

For me, mystical experience means feeling strongly the presence of God. And this has happened a few times for me. For instance, I once went on a trip that involved a long period of being out in nature by myself. I had been looking for bald eagles. In the course of this, the world opened up.

It was a bright, sunny, windy winter day. My senses were heightened with the feel of the wind, the reflection of the water, and the smell of the salt marsh. I had an acute sense of being a part of everything. It was an instant that didn't last long, but it was beyond time. It was endless.

Sometimes I think about what it's like in the desert looking at stars, an upside down bowl of stars. I told my cousin that, and she thought I was crazy. I was so taken aback. It was so special for me, and she didn't get it at all. This Christmas, she took a trip on the Amazon and talked about looking at the stars. She apologized to me; now she knows.

Judith's Knowing: 'A place inside me'

In the early 1980s, I experienced a life change.

There was upheaval with my kids, who were in their teens. Another relationship was breaking up after divorce. I was facing lots of unknowns and feeling that life was out-of-control.

I went to a conference. Yoga, meditation, and high-intensity sound were used to awaken the *chakras*, the body's spinning energy centers.

Creative dance and other rituals around healing were very opening for me. When doing hands-on healing, I knew deep within (not from the mind) what that person needed. I was more free inside and expressive to others. I showed more love and compassion for others from a very centered place and the experience stayed with me in the ordinary world.

My kids asked what was going on. I was so different. The teaching I learned at the conference was said to embrace the shadow side when it comes so that you can heal. When I think back, the 'knowing' went on for a long time. I remember that I walked into a bookstore once and knew what was in the books. I felt a compassionate caring for strangers, as if I knew them. I also felt deeply drawn to know more about Jesus. My whole life changed.

Then I began coming down. I began going into the depths of me, into the dark unknown. As I began to be enveloped by darkness, I got very afraid; the depths were as deep as the heights had been. My mind wondered, 'Am I crazy?' But I instinctively knew I wasn't.

For a couple of years, I was in a state of confusion: in and out of a shadow. By 1984, the darkness was so great that I became suicidal. I felt the pain and anguish of the world. The inner pain became intolerable. I thought about death; the pain was too profound, too much pain. However, in the end, knowing that I had the choice to end my life was empowering. I consciously decided not to. I decided that I wanted to see my kids grow up. I decided that if I'm gonna live, I'm gonna find out what it really means to live.

A place was touched inside me and outside me and around me — a reality more real than lightness and darkness. An all-knowing, all-powerful presence told me I wasn't crazy.

One day I was asking myself: What is next for me? What do I need? I felt attached and attracted and wanted to be back in this mystical place. In the course of this, I saw a large place on a rolling hill with lots of people in it. Later, when I went to a yoga and retreat center in Massachusetts, I saw that it was the very place that I had imaged.

For ten years, this retreat center became my spiritual home. I realized that a place inside me was guiding me all along.

Jane's Knowing: 'Deeper than I thought possible'

For me, mysticism is a searching for direct guidance from God, not mediated through any other person or through Scripture.

I myself have had mystical experiences while in Quaker meeting. One was close to the 'white hot center' of meeting. I don't know what the question was, but I remembering answering, 'OK, I'll do it — but this time let the healing be complete!'

That was a sacred moment. At the time, I had no memory of abuse. But it emerged, and it was overwhelming. There were intense consequences to my praying this prayer for complete healing. When I said, 'Let it be complete', I was asking for knowledge deeper than anything I could then imagine.

That sacred, God-given moment had consequences. The work that I needed to do started. I was in crisis for about three years, but that crisis was part of the healing. Even when I didn't feel the Spirit, I knew it was there. I don't think I could have done it without the

Spirit, without medical intervention, without poetry, and without (F/f)riends. I was fortunate that resources were provided. The process is on-going.

The trigger for bi-polar disorder went from baseline to psychotic in forty-eight hours or less. The break happened on a Wednesday night. I slept, and then I spent the next day psychotic. My husband and son left for the day. My delusion was that I was to arrange universal peace. I ritually rearranged the furniture in the house to create patterns. Then my husband and son came home. I couldn't stop. I couldn't even eat. I kept trying to create patterns. I got very 'not gentle'. I attacked the police when they came. I was obsessed with reconciling opposites. This was not like me.

I was given antipsychotic drugs, and I slept, but much of that time is not clear in my memory.

I know it was psychotic because when I look back, I say, "That was not normal." But when I was going through it, it was very real. The episode was quite damaging to my concentration, and it still takes a while to reorganize.

In seeking a healing that is complete, I have been strengthened by many experiences and many gifts. And those gifts have led me to a knowing of myself that is deeper than I ever thought possible.

Jean's Knowing: 'As if in another dimension'

In 1984, at the age of 46, I was opened to experiences of which I had no prior concept. The changes in my body were perceptual, visual, auditory, and olfactory. These transient perceptual changes occurred when my body, although physically still, was experiencing tremendous energy. Every cell felt highly charged and aware.

Visually, I had the ability to see detail on a mountainside at a distance of a mile as if I had high-powered binoculars. This could last five to eight minutes. Walking into a store, I could see a glow around all the clerks and customers. Time, as I had known it, sometimes seemed suspended, and I would feel as if I were in another dimension. I could hear sounds and occasional words and knew these were occurring in this other dimension.

At one time, I seemed to be part of historical events, such as the Crucifixion or the Holocaust, all of which seemed to be taking place in the present moment. At times, I enjoyed the smells of sweet perfumes, freshly baked chocolate chip cookies, or baby powder, and these smells would be present even when they had no outer explanation.

There were also thought-process changes, changes in feelings, and behavioral changes. I was experiencing tremendous love and power for which I had no preparation or explanation. I also had a new and strong sense of inner guidance that gave me the strength to accept what was happening to me. I saw it as an adventure — and an unfolding story.

The explanation that fits my experience is the 'Kundalini awakening' that is described in the Hindu tradition. Early Christians referred to it as the Holy Spirit or the Holy Terror because of its unpredictable effects. The process leading to the quickening of this energy has been understood by mystery religions through the ages and has been the goal of yogic practices.

My husband (ex-husband now) was a medical doctor, and he was able to listen to my experiences for a couple of days. Since the only understanding he had of such events was in his psychiatric training, he felt I was experiencing some type of mental illness. At the urging of a colleague, he persuaded me to see a psychiatrist. I knew I was having a spiritual experience. But as a nurse myself, and thus a part of the medical establishment, I knew there was no way I could convince anyone of this.

I was able to cope with the changes because of the sense of a loving guidance and support within myself. I managed to get through difficult times by taking one minute at a time and not looking too far ahead to the future.

Marshall's Knowing: 'Listening to my Inner Guide'

For me, mysticism is simply a name that has been given to that direct invasion of our feelings by the divine Presence. I believe that we all possess an inner faculty to respond to this direct invasion, which leads to a greater knowing of the spiritual world.

What happens is beyond words. It is experiential, ineffable. It is not extrasensory perception, and it should not be associated with occult experiences. It is not confined to special temperaments. It is not an escape from the realities of this everyday world.

Sometimes it is very individual. But it can also be a group experience, such as might happen during a Quaker meeting for worship based on silent waiting and obedience.

Jesus was a Jewish mystic, and it seems to me that the early Christian experience was a mystical one, an experiencing of the Inward Christ. Those who have experienced a 'gathered meeting' for worship know the power of that experience, which is mystical.

A mystical experience can be transforming. It can change one's life. It is a fleeting experience, but a *knowledge-giving* experience. I believe it grows out of an inward faculty that can be cultivated by spiritual disciplines — regular attendance at a Friends meeting for worship, for example — or by awareness of being on a spiritual journey.

On a personal level, when I was an intern teacher at a Friends school, facing the prospect of contacting my draft board, I struggled to find a direction. I was divided in mind and spirit, not sleeping nights. I felt alone in the midst of friends. Few were aware of my inward struggles. I even felt my loving parents would not understand. This lasted for weeks until I could at last know an indescribable wholeness and inner peace. That wholeness and peace grew in me day by day. I felt in touch with the world around me in a fresh and energy-giving way.

It followed, of course, that I faced my draft board and became a conscientious objector. I now look back upon this illumined experience as an opening in my life that set a course for a lifetime. It was not an intellectual experience. It was an ethical mystical experience.

Mysticism is not a retreat from the world. It is the other side of action in the world. All this is in the Christian context, but it is inclusive. I came to it through Christ.

Sonia's Knowing: 'At the core of reality'

I see mysticism as the quest toward the 'Great Unseen', that presence, that being-ness that is at the core of our existence and at the core of all that we know as 'reality'. It's a movement toward that which unifies and underlies all, a movement toward 'the All-in-All', as Mary Baker Eddy might say.

I have had many mystical experiences in the course of my life. Perhaps the most powerful has been my quest to understand the nature of spiritual healing.

Between 1990 and the present, I have conducted over four hundred healing sessions with individuals. These sessions have been highly mystical and appear to draw upon a higher, unseen power.

At the time the 'work' of healing began, I did, indeed, fear that I was going crazy. I wasn't fearful as much as 'pissed'. I felt deep ambivalence about being drawn into this force field, and in the early days I often questioned how this 'might look' to other people, what they might think about it. I questioned what I thought about it, too.

The healing sessions demand a kind of 'extra-sensory perception', I suppose. It feels like 'feeling from the belly' when I work with another person. I 'read' their energy field and apply 'toning' — human sound — to their bodies.

But I see the mystical experience as far more comprehensive and dynamic than simply perceiving something that is "extra" ordinary. It feels more to me like a huge tapestry of experiences, a deep or high dimensionality. It's almost as much being communicated to by a higher force as it is communicating with that extra-dimension that is beyond typical or ordinary 'reality'.

Abbie's Knowing: 'The rational and the intuitive'

As a teen, I had a favorite lane, a playground. I took long walks by myself and needed solitude. In my twenties, I worked at a camp and resisted moving to the city. I didn't want to lose my solitude.

One day, I went out by a waterfall and fell asleep. I was awakened by the same kind of light experience that I had had as a child, an enormous light that comforted me. Immediately, I *knew* which direction to go in a decision I had been pondering, even though it meant I was going to miss my friends.

I shared that incident for the first time with a teacher at Pendle Hill, a Quaker retreat center. I shared it because in the lives of Teresa of Avila and Julian of Norwich, I had found the language to describe my experience of knowing.

In my marriage, I developed intuitive communication as well as the rational. And I found that the more I acknowledged intuitive capacities, the more I became pre-cognitive — and knew things before they happened. At first, it was scary. Now it is OK. I am used to it.

That which is rational is taught; the intuitive is not. Acknowledging both allows the spiritual dimension to flourish.

At Pendle Hill, my reading has provided me with a reference point as to where blockage is to my intuitive knowing. The mystical aspect of my life had not been integrated into the rest of my life. Until ten years ago, I had lived two parallel journeys. I needed integration.

Trish's Knowing: 'A gift of grace'

There are times when I know and experience the direct presence of God.

Sometimes it is just an overwhelming sense of God's presence, a totally focused attention with everything else dropping away. Other times, it's hearing words, coming to me as 'loud thoughts' (a phrase coined by Gerald Priestland, an English Friend), not originating from me.

One Sunday in 1985 there was a little Quaker gathering in our town. One member lived there, and Quakers from all over were traveling to her home for a meeting. I had been clinically depressed for some months. During that meeting, I said silently, 'Are you there, God?' and God said, 'Yes.' After a while, I said, 'What about these troubles I'm having?' and the answer was 'These are the sorts of problems that can be overcome.' This experience affected me profoundly, and I spoke in the meeting, telling of what had happened. The fact that I had spoken in meeting seemed to fix it in reality.

I continued to have mystical experiences, and sometimes, these were not comfortable. I told a few religious friends, who were fairly dismissive. I was still extremely depressed and wondered if I were going mad. My psychologist told me that I wasn't, but I didn't believe her. She eventually became quite angry with me and refused to see me any more.

I was distraught and eventually saw a psychiatrist, who prescribed medication to relieve the depression. He asked me if I was hearing voices. I said no, for I knew that what I was experiencing was not schizophrenia. Still, these continuing experiences caused me to question my sanity.

I didn't feel reassured until some years later when a weighty Friend spoke with me and asked: 'Are the messages in the context of worship?' 'Do the messages lead to good?' 'What are the fruits?' Gradually, I began to take my experiences for granted as a spiritual reality.

Mystical Knowing: Overcoming the Roadblocks

The stuff of the world is there to be made into images that become for us tabernacles of spirituality and containers of mystery. If we don't allow soul its place in our lives, we are forced to encounter these mysteries in fetishes and symptoms, which in a sense are pathological art forms ... (Thomas Moore)[4]

As Thomas Moore suggests, when the road to spiritual knowing is blocked, pathologies are created. These pathologies result from thwarting the mystical energy which a person is being given or from labeling a person's spiritual experiences as a dangerous illness when they could more productively be framed as visionary gifts.

Sometimes even those who have been given a degree of mystical knowing find themselves becoming afraid or turning back. Many of those whose stories I have heard experienced what they call 'a time of darkness'. This experience is so universal that Evelyn Underhill, in her classic book *Mysticism*[5], suggested that the third stage in a mystical life is 'the dark night journey'.

Although this stage in the process can be frightening and confusing, and although some have called it a wrestling with the 'daimonic', Underhill argued that it is a necessary part of the journey. Only by experiencing the dark, she wrote, can the Light shine forth.

M.C. Richards, an artist and writer on the mystical journey, put it this way:

> We choose to wrestle with the daimonic because it will increase our self-knowledge, it will help our growth and creativity, and it will enrich our lives and capacity for 'good'. Our picture of life will change. Wholeness will become a suffering-through of both dark and light. And the meaning of 'Judge not!' will become clearer. Truth will include not only the shadow but the dark 'unknowing' out of which consciousness and creativity come. The dark in both senses is the source: dark as the 'unacceptable' and dark as the intuitive. Our sense of fact expands. Our epistemology will develop to include consciously the realms of 'spirit knowing': imagination, inspiration, intuition. In tune with this, life is asking us to develop a language true to the facts. This means that we may sacrifice our one-sidedly intellectual language for a language more adequate to the resonances of the double realm ... An inner Man stands as an arch between the polarities of being and transforms them, integrates them. He rides and guides the fire.
>
> To see in the dark takes new spirit, new eyes. It is difficult to work to contain reality and not to falsify it. We need to be both vulnerable and unwobbling if we are to be open to contacts with the spirit world, both dark and light. But when we can do this, each 'receiving' and 'offering' helps to befriend the realm, helps us to be able to say both 'yes' and 'no' to it. And we ourselves are befriended.
>
> I had a dream in which a tremendous fire was raging clear across the full width of the horizon. It burned steadily. We ran away, though one person remained. The fire swept through the landscape,

---

[4]   Moore (1992).
[5]   Underhill (1911).

through the house, through the pottery vessels, through the man. I could see the flames coursing through everything, but nothing was being consumed. After the fire had swept through, I returned and the man said, 'Everything is still here — only the color is deepened.' And it was so. He was intact, but the pots were richer, deeper, and more lustrous in their colors.[6]

I am reminded of M.C. Richard's words every time I recall the first conversation that I had with a seventeen-year-old male, who I'll call Steven.

'I am full of energy, and it has turned to fire,' Steven told me. 'You had better be careful. My fire sends out comets and burns people. I don't want to hurt people, but I can't control the energy. It feels like the devil gets inside of me. It is a fire out-of-control.'

I asked him what the fire meant. 'It means hell, the devil, anger, madness, rage,' he said.

'Is there anything good in your fire?' I asked him. 'No!' he replied. 'Some fire might provide heat or light, but the devil has gotten in me and taken over. There is no way for my fire to do good.'

We talked about how fire fighters put water on a fire to put it out. But Steven didn't want water put on his 'fire'.

'Steven,' I said, 'maybe we don't need to put out your fire. Maybe that fire is your life's energy. Maybe we just need to contain it so that it is not out-of-control. Then maybe you could learn to direct it creatively rather than destructively.'

'I really want to,' he replied, 'but I can't.'

Commenting on his description of the fire as 'the devil', I asked Steven if he was religious.

'I haven't been going to church lately,' he said, 'but I used to go. God has always been — and still is — an important part of my life. But now I have to ask, "Are you there, God, or am I going crazy?"'

Until the tenth grade, Steven had been a steady, straight-A student. Then the 'fire' had come. He became aggressive toward himself and others. Despite the gravity of the situation, I saw a spirit in Steven that was beautiful and brilliant.

Steven was being opened to a new level of consciousness. Like many of us, when he was opened to a new level of consciousness and God, he was also opened to his unconscious, with unresolved trauma. That brought the fear of nonbeing, based on the actuality of some part of him facing nonbeing. For Steven, it was a deep trauma involved with facing the loss of his own ethnicity and culture. He had been adopted at birth from Latin America by white American parents who loved and cared for him. Emotionally, he had tried all his life to please others, to be 'white' in every way he could — and he had failed. Although he had tried to do everything 'right', prejudiced students at his school had rejected him. He was in despair.

In his old life, he had tried to be accepted by taking on the values and dreams of others. That life had cracked. New life was necessary. He would find the Light only by moving through the 'dark night' of the soul. He would come to a deeper level of knowing only by passing through an anguished time of unknowing.

I empathized with this young man because in my own life, I, too, have experienced periods of chaos. As a youngster, I was very shy. Much of the work in my life has been about finding my own voice in a world that did not always seem to value what I had to say. The image for me is of a little bird learning to sing out its songs. And each time God calls me to a greater awareness of a bigger universe, some darkness and chaos seem necessary. An old life is grieved. A new life bursts forth. Then peace returns.

For each of us, there are multiple levels of knowing and unknowing. Some years ago, in thinking about those multiple levels, I adapted the insights attributed to an unknown Jewish rabbi to describe four different lenses through which we can experience a degree of 'knowing'.

The first is the physical lens, the lens through which we gain knowledge of the world of concrete facts and material objects. Our Western culture focuses primarily here. Our values in this plane center around material possessions and the related status they give us among our peers.

The second lens opens us to the metaphorical world, which provides us with rich symbols. Through this lens, we begin to gain some knowing of the world of the Spirit, and we become aware of the power of our unconscious minds. As you may have noticed, my first conversation with Steven took place primarily in metaphor.

The third lens brings us in touch with the archetypal world. Carl Jung taught that this is where our personal stories begin to overlap and intersect with that which is universal. Fire, water, the struggle with good and evil — these are archetypal concepts. This struggle with something bigger than ourselves has gone on since the beginning of human life. Steven was there when he felt something invading his body that did not seem to be coming from himself. Many people get glimpses of the archetypal world at some point in their lives, some sense of knowing.

The fourth lens leads us to the kind of knowing that is possible only in union with God. Some people occasionally get brief glimpses that leave them longing to live life with this kind of knowing. But living in union with God while on the physical earth is not easy, and only a few are graced with more than a glimpse.

When we live our lives according to the norms of our culture, we know only at the first level. We care deeply about our physical possessions and about our acceptance by others. Some people, locked into this more limited knowing, even go so far as to determine others' worth by their social status or physical accumulations.

Our culture tends to define life as 'normal' when it is lived on the physical level, with perhaps occasional glimpses into that which is metaphorical. But some people are called to know at a deeper level. Such knowing isn't always easy. When we move into those deeper levels, we may find that our life feels chaotic, especially when we hit up against the norms of a culture that expects us to know only through the physical lens of concrete facts and material objects.

Through my own experience, however, and through my conversations with others, I have come to better understand these periods of 'dark chaos'. I see them now as sacred gifts. The metaphorical 'darkness' that they bring is warm — and good when we can let go of fear. When I view it through the archetypal lens, I see it as the condition of the universe before the universe was created. The energy that comes is creation energy. When we are called to deeper levels of life with God — and greater use of our creation energy — we must enter this dark chaos for a time. From that dark and sacred chaos comes new life.

Everything is temporary, including the chaos. We do not stay in that place of deepest despair where our fears take us. We emerge stronger, fuller, endowed with a deeper level of knowing, more able to share our knowing with others.

In the 1970s, John Weir Perry, a Jungian psychiatrist, experimented with a residential program in California that he called Diabasis. He found that most people who had psychotic breakdowns got through the episode in about six weeks and most never needed to be hospitalized again. When not medicated, most found their experiences to be transformative spiritual encounters, moving them into greater understanding, deeper knowing, and richer lives. They became 'weller than well'.

In his books[7] Dr. Perry argued that psychotic episodes could more accurately be understood as a natural psychological or spiritual healing process. Sometimes the energies that come through this kind of divine gift feel good; sometimes they feel bad. But the energies themselves are neither good nor bad. Energy is just energy. *What we choose to do with the energy* is what makes it effective or ineffective in our lives. At first, we may not realize it, but the energy that comes is usually right for the tasks we have to do. However, there are days when there isn't enough

---

[7]   Perry (1974, 1999).

energy. And then, for some of us, a storm hits, and there is energy beyond what we can manage and use constructively.

Earlier in this chapter, I shared the stories of those who used various religious traditions to navigate the storms: Jean, who found, once she had a comfortable framework for understanding the 'knowing' that she had experienced, the fear was relieved — and she could integrate that knowing into her physical life; Jane, who used her Quaker religious tradition and appropriate psychiatric frameworks to move toward integration of the 'knowing' she had experienced. Finding useful elements with the tradition that best suits the experiencer is important. Steven was Catholic, and the lives of the Catholic saints and mystics would have provided him with nurture and words for framing his experiences, if only he had known about them.

Ironically, many professionals are often the least qualified to be helpful when people are experiencing new ways of knowing. All too often, their response is either 'diagnosis and treatment' or a callous dismissal of even the possibility of a deeper level of knowing.

One night Billy showed up for a presentation I was giving. He told me that he was twenty eight, had been diagnosed with paranoid schizophrenia, and still lived with his parents. He said that as a young teenager, he had heard a voice that he believed to be God telling him that he was to be a monk. He talked with his priest about becoming a monk, but his priest told him that God does not talk to people in this day and age. Such knowing happened in Bible times, said the priest, but direct communication between people and God ended centuries ago. By the time he spoke with me, he was convinced that if he had listened to his calling from God — if he had accepted the knowing that he had been given — he would not be in the predicament in which he then found himself.

To dismiss deeper levels of knowing is not only callous. It's dangerous. Consider Thomas, a man who is in prison for murder. As I listened to his story, what I heard was that he had had many experiences of the Divine when he was young, experiences that he did not understand. He wished there had been someone who could have heard his experiences and helped him understand them and know how to live with them. He was sure now that there was a creative component that could have been directed in a different way. Instead, his despair led him down a destructive path.

By thwarting the creative and by unnecessarily labeling the visionary as pathological, our world creates pathology and loses the creative gifts and constructive growth that seek to emerge from a richer level of knowing. Again and again, we cut ourselves off from diverse kinds of knowing. Again and again, we strip Spirit from our vocabulary.

The mystical way of knowing opens us up to the largeness of the universe around us. But all too often, powerful forces close us down. Mystical knowing makes us bigger. Mystical forms of knowing lead us to lives committed to cooperation, love, kindness, and those other virtues that climb high and go low to the seed of foundational building blocks. The inner life with God fuels an outer life of knowing service.

There are social obstacles to this kind of knowing, but I have found that many have learned to overcome these obstacles, taking hold of deep, alternative lenses that greatly enrich our world.

Section 2

The Perspective of Psychology

While the breadth of the vision just presented is a product of our times, readers will recognise elements of it from the teachings of spiritual and social visionaries over the past two millennia. We often feel despair that apparently little of this teaching has taken root, and wonder if there is a fundamental reason, something basic to the make-up the human being, for why humanity seems so impervious to what many regard as the obvious. To address this, we need to examine the knower, the nature of the human person.

In his chapter, the neuropsychologist Douglas Watt focuses on the central role of our emotions. For much of the human race through much of our history, fear and/or rage have grabbed the cognitive and emotional workspaces making spiritual learning something of a luxury; and when spiritual teaching has received attention, our emotional makeup has distorted it into the structures of religion, in the West usually based on an anthropomorphic concept of God. The driving forces for both religion and mysticism, he argues, are our underlying attachment mechanisms. Attachment is seen as a biological mandate for hominid brains, the source of our deepest comforts and joys, the loss of which drives our deepest pains and sorrows. Reverence and awe, as a finite if powerful hominid brain confronts an infinite natural world, are argued to be the affective core of spirituality. Those deeply interested in spiritual perspectives have throughout the ages been often torn between deep hope and equally deep worry. This perhaps has never been more true, given that we are now perched on a precipice of an unprecedented ecological disaster reflective of the deep failure of traditional faiths in a technological age in which nature is seen as an 'object' to be manipulated and mastered instead of 'the ground of being'.

The consequence of this is a frightening rate of increasing exspeciation and impending loss of vast biological diversity, driven in part, Watt argues, by harsh in group/out group distinctions that human beings seem to excel at, a tendency mirrored in and reinforced by religious 'sect-ism'. Deeper appreciation for the underlying affective themes in religious searching, as distinct from the current much more divisive focus on the cognitive forms, is seen as one potential antidote, and in taking this way forward, a science that is revealing the world in terms of a hierarchy of emergent properties can unite with a spirituality freed from anthropomorphic notions of God.

It is noteworthy that, on this approach which stresses the role of emotion, no claim is made by Watt that any aspect of religion *is* a way of knowing in the cognitive sense, and in the case of the popular religious beliefs on which Watt concentrates he argues that these beliefs are definitely not (valid) ways of knowing. Nonetheless, in practical terms his conclusions align closely with those of other writers in this volume.

Many approaches to the human predicament see us as composed of conflicting parts. Indeed, any scheme involving different ways of knowing naturally suggests a division of the human being into different parts, or at least different faculties, corresponding to these different ways. Plato stands out as an expositor, with his graphic depictions in the Republic and the Phaedrus of the division of the human into body and soul, as separable components. There is also a tradition of the parts of the human fitting badly together, or being at odds with each other. St. Paul was famously frustrated by the way that 'the good that I would I do not: but the evil which I would not, that I do ... I see another law in my members, warring against the law of my mind'.[1]

Isabel Clarke, a clinical psychologist, uses this concept of misfitting parts to offer a complementary approach to that of Watt. Like him, she stresses the vital role of emotional and sensual components of the mind of the human being, while also drawing on research into memory and information processing. She sees the functional separation of the two principal components of the mind as like a gulf in our inner landscape. One side is represented by the mystical quality of experience which is the subject of much of this volume. The other side is represented by our rational faculties, so that the more familiar quality of everyday experience manifests when the two sides of the gulf are well connected. These two ways of operating give us the basis for two ways of knowing: one cool, analytical and logical in the conventional sense; and the other which is wonderful, paradoxical, relational and without clear boundaries. In terminology modified from the 'interacting cognitive subsystems' (ICS) model of Teasdale and Barnard, she terms these two the *propositional* and the *relational* subsystems.

---

[1]   Romans 7.19, 23.

Our society tries to ignore the relationally based way of knowing in its elevation of mechanistic science and technology, or to harness it to the needs of the market by introducing it into the imagery of advertising and alcohol. But this way of knowing knows no such restraints. It seeps back in fundamentalism, drugs and cults if it is not embraced in more wholesome ways.

This model of a twofold mind, with a strong connectivity between the two components, sheds light on many of the issues raised here. Clearly many of June Boyce-Tillman's polarities (such as 'rational/intuitive') would derive from these two subsystems. More generally, it gives an extra dimension to the cultural history of the domination of feminine ways of knowing by patriarchal ways of knowing. This can now be seen as one of progressive political domination by an emphasis on the propositional subsystem, with its attendant tendency to separate from and control the environment, to the detriment of mysticism in religion and of indigenous and feminine spirituality in society, with their tendency to strive for deeper connections with the environment, and a more internalised understanding of it.

It cannot be stressed too much that, because all our meaning-making flows through these two subsystems, they determine our universe, in so far as it is knowable by any form of thought.[2] We are thus dealing with a much more fundamental level than, say, the more publicised 'right-brain/left-brain' division,[3] which refers to different balances of ability in handling the (given) universe.

---

[2]   The only qualification in this might be the highest mystical experiences marked by a form of knowing-by-being.

[3]   The strongest counter-argument to this is perhaps Shlain (1998). In fact it is likely that there is some correlation between left/right brain physiology and the implicational/propositional division, just as there is correlation with old/new brain physiology. ICS research identifies functionally independent subsystems which are almost certainly implemented in a way that cuts across the anatomical divisions.

Douglas Watt

Attachment Mechanisms and the Bridging of Science and Religion

The Challenges of Anthropomorphism and Sect-ism

That which sings and contemplates in you is still dwelling within the bounds of that first moment that scattered the stars into space. (Kahlil Gibran)[1]

Modern man is estranged from being, from his own being, from the being of other creatures in the world, from transcendent being. He has lost something – what, he doesn't know; he only knows that he is sick unto death with the loss of it. (Walker Percy)

Everything that the human race has done and thought is concerned with the satisfaction of deeply felt needs and the assuagement of pain. One has to keep this constantly in mind . . . Feeling and longing are the motive force behind all human endeavor and human creation, in however exalted a guise the latter may present themselves to us. (Albert Einstein)[2]

Introduction

This chapter attempts to connect science and religion by critically dismantling fundamentalist notions about faith as anthropomorphic.

[1]   Gibran (1991), p. 83.
[2]   Einstein (1949), p. 24.

Anthropomorphic notions of God as a person obscure potential bridges between mysticism and scientific conceptions about nature as a recursive hierarchy of emergent properties, but anthropomorphisms also provide insights into the underlying attachment mechanisms informing much of religious searching. Attachment is seen as a biological mandate for hominid brains, the source of our deepest comforts and joys, and the loss of which drives our deepest pains and sorrows. Comparative religious studies have often times been hampered by attention primarily to the cognitive forms of various religions, with relatively little attention to these underlying affective themes. Reverence and awe, as a finite if powerful hominid brain confronts an infinite natural world, are argued to be the affective core of spirituality. A reverence for the mysteries of nature and appreciation for what science is now revealing in terms of a hierarchy of emergent properties (as opposed to positivistic scientism) are argued to be deeply compatible with the core of religious mysticism, emphasizing the 'oneness of all things'.

Affective Issues Embedded in Religious Searching: The Mandates of Attachment

In the past three decades, emotion has gone from its scorned position within the heyday of behaviorist doctrines as irrelevant, to an exploding topic within many branches of psychology and neuroscience. In this chapter, we will be building bridges between a proposed ontology of science and an analysis of core emotional aspects of religious mysticism and belief. An organizing hypothesis is that *at the apex of the range of emotional quests and strivings organized in the human mind/brain sit inescapable and utterly compelling religious questions and dilemmas*. Indeed, these religious and spiritual issues exert a fundamental organizing effect on many other affective phenomena, in some sense sitting as a 'supervisor' over most if not all other values and affective ideas. Rather than being 'outside of science proper' (their traditional framing in our culture), religious beliefs instead will be explored here as a critical and largely neglected sub-domain within general emotion theory. This framing is clearly not the traditional approach to understanding religion, which has more traditionally assumed that scientific perspectives on emotion and formal systems of religious belief have little to say to each other. Most of us in Western cultures subscribe to the notion that questions about purpose and meaning in life lie outside the domain of science. An alternative position would be that that science can and indeed should examine emotion as the wellspring for all value in the mind/brain, including especially our most emotionally resonant ideas, those found in our many religious and spiritual traditions. Thus, our discussion must be of necessity deeply inter-disciplinary.

In this chapter, I will outline a potential bridge between the two traditionally antagonistic domains of science and religion derived from an analysis of the deep structure of attachment. Attachment is a central biological mandate for hominid brains, indeed, it may be the unappreciated developmental cost for, and functional 'flip-side' of, our large cortex which requires our subsequent long period of infantile dependence and relative helplessness. Attachment and its many derivatives are the sources of our deepest joys and comforts, and traumatic loss or severe dysfunction in a primary attachment potentially plunges us into perhaps the deepest pains that humans can experience. The point of this chapter is not to address the objective truth or falsehood of various beliefs systems, but instead what they might mean in more fundamental affective terms, what in some circles has been referred to as a 'hermeneutic' or 'iconic' approach. I will argue that emotions neurodevelopmentally grounded in basic mammalian attachment mechanisms are the source of the real (and undeniably great) power of religious ideas. This is not a 'debunking' or any version of a reductionist argument, rather it points us more towards their real and primal sources of their profound power.

The traditional antagonism and hostility between science and religion is driven by two primary issues: (1) a traditionally 'anthropomorphic' framing of religious issues in most Western religions (God is a 'Daddy-in-the-sky' figure); (2) a rigidly ideological and positivistic 'scientism' on the other hand (science just reveals emotionally empty mechanisms and nothing more). This dichotomy is badly outdated, grounded in conflicts that are centuries-old. Instead, I would offer an hypothesis (one Einstein intuited clearly more than 50 years ago) that reverence and mystical awe in the face of nature, and the intense curiosity these affective states engender is an important and neglected foundation for all the sciences, and all the best that flows from them. Organized religion and systematic theology may also have their most important roots in just such mystical reverence and awe. However, organized religions throughout human history have often demonstrated 'selling out' (or at least deeply obscuring) that foundational component, mostly to their great detriment. This has often taken place in the service of promoting and often 'calcifying' religious doctrine, typically in the interests of promoting social-political power and influence. This has often been tragically associated with harsh, even violent, in-group out-group distinctions based on equally harsh dichotomies of belief. Thus, organized religion has been intrinsically linked to some of the worst atrocities that humans have visited on one another, as well as being closely related to our deepest spiritual values. This is less paradoxical if one considers that religious beliefs 'refract' (rather than independently create) virtually all major emotional trends in humans, from the noblest to the most savage.

Of course, many, many writers in many disciplines have commented on how the driving forces behind all religious beliefs systems are profoundly affective issues: 'why am I here, what is the purpose for my life, what are the ultimate realities around and inside of me, and how and why was I created'. These are not simply cognitive challenges, but are among the most emotionally loaded questions human beings can ask themselves. Less appreciated is that all of these issues are fundamentally cognitive extensions of needs stemming from systems in our brains organizing primary attachment: we are seeking connection and continuity, in the face of abundant evidence that life offers these things at best rather inconsistently and quite intermittently. Even the greatest and longest loves have final endings, in that one partner must die before the other, and our adorable little children grow up and leave us. And for many, love and connection is in desperately short supply throughout life anyway, such that basic longings for some guarantee of love and connection are a powerful driver in much if not most religious searching. Thus, religious issues and spiritual searching cannot be adequately understood without outlining their fundamental connections to brain attachment mechanisms, which are thought to reflect an evolutionary extension of basic homeostatic mechanisms to preserve the body that developed strongly in early mammalian lines. These social attachment systems are further elaborated in primate and hominid lines, and aside from organism defense (prototypically expressed in states of fear and rage) and basic homeostatic mandates (thirst, hunger, sleep etc.), are our primary sources of value in life. Yet attachment for all its importance is still quite poorly mapped and even neglected in much of current neuroscience.[3]

However, for those impressed with our cognitive abilities, our extended cortical capacities and our prolonged period of infantile and childhood emotional dependence are two sides of a coin, rarely seen for their intrinsic tie to each other. The more cortical the creature, the longer the period of infantile dependence required to insure cognitive learning and motor mastery, and thus the more profoundly important attachment processes must be to that brain. Thus, our higher cognitive abilities, and our profound emotional vulnerability are much more intrinsically connected than most realize. This appears poorly recognized in sections of our culture that value intellect and cognition in

---

[3]   See Panksepp (1998) for summary.

terms of its having some putative separation from emotion (where none, at least in any absolute sense, appears to exist). There is instead much evidence[4] that we are born out of an early attachment matrix as separate conscious creatures, and that the whole course of our neurodevelopment is profoundly colored by the sustaining quality or failure in those early attachments.

As many have commented, including Einstein, religious searching is also deeply motivated by our existential awareness of our biological and psychological vulnerability and most of all (at the apex of human vulnerability), our mortality. When one combines the social grounding of our brains with a potential awareness of mortality, and the manner in which images, myths and legends resonate primarily with affective abilities of our right hemispheres, *these three intersecting processes jointly define a rich and fertile ground out of which religious and spiritual seeking can grow.* Through the ages, our deepest comforts and reassurance have often come from countless myths, images and stories since the dawn of culture that affirm (in some fashion) that death doesn't triumph over all we have built and loved, and that we are a part of something enduring and permanent. *In this basic and profoundly important way, religious faith is what psychoanalysis has called "object constancy", only now writ large: a complex mythic-symbolic umbrella sitting over the early internalization of (hopefully) good enough parenting.* A critical component of many religious beliefs is the longing for connection and comfort, comforts that early in life are only available from a loving and devoted other, but that we later seek 'abstracted' in religion. As Martin Buber stated in his famous work *I and Thou*, God is an eternal *Thou*. Religion seeks no less than an emotional tie in which we can never be abused or deserted, a relationship beyond harm, beyond the many basic human failings in our all too limited and flawed human attachments. This proves to be a very tall order, as those themes and every associated human vulnerability follow us doggedly into every corner of our mythic and religious constructions. This becomes another piece of compelling evidence that attachment provides many of the essential templates for the construction of religious belief systems.

Religious Beliefs as a 'Theological Rorschach'

There are many ways to address this tug of war between faith and doubt, between our wishes and our fears. Two of the most common are: (1) pathological certainty and 'doctrinal rigidity' (if one seeks in a sense to protect faith from doubt); or (2) by giving up the quest altogether, if one feels that no faith can really be abiding or enduring — that all

[4]   See, Schore (1994).

religion is ultimately a sham, as Freud said, a narcotic illusion. Neither of these is a great answer in my judgment, but they both surely exist in abundance. The first position might be thought of as expressed in virtually all fundamentalist sects, or any religious belief system claiming to be the one absolute true-and-only-faith. The second position is visible in much of middle-class secular atheism, and in positivistic-reductionistic interpretations of science. Because of the intrinsic tie between attachment mechanisms and religious beliefs, the conflicts embedded in religious faith and disillusionment are no different, in terms of their fundamental structure, from the day-to-day struggle that we all have with our ambivalence about other people. We live in the ebb and flow between efforts to protect our idealizations of others (and of ourselves) versus struggling with the sequellae of the collapse of those idealizations, since neither we ourselves nor other people are as good as we might need or wish. At moments of most grave crisis and trauma, the collapse of our idealizations of others and ourselves is nearly total. At those points there is intense personal suffering and anguish, though some learn from this process and become stronger and more resilient, while for some individuals, trauma only deepens bitterness and despair, and is experienced as part of seemingly endless enactments of childhood traumas.

Psychoanalytic object relations theory outlines how master templates or images of all social events are constructed from internalization of early, affectively charged interactions with parenting figures.[5] If one looks closely, religious images, and those deepest internalized images of self and other are inseparable, *as they interpenetrate one another and are ultimately woven from the same cloth.* Whether we think we are interested in religious issues or not, we all struggle to read the inscrutable face of the Cosmos around us, and we see (to some extent) what parents and other loved ones have given us. In this sense, *religion is no more or less than a great 'theological Rorschach'*, revealing the most profound aspects of ourselves and what we have internalized. This 'projective' aspect of religious belief is intrinsic to the manner in which basic attachment mechanisms inform religious searching and belief. In clinical work, I have frequently found that the patient's image(s) of God, whatever those might be, shed a very penetrating light on that person's deepest images of self and others, born out of formative early family experiences in their primary attachments. I have often seen depressed and despairing patients in crisis admit to feeling that God, (like everyone else in their lives), must have abandoned them. Given the current state of the world, the notion that God has abandoned us is easy to empathize

---

[5]   See Watt (1990) for summary.

with. Inherent in our image of God as some kind of magical rescuer is that we must inevitably feel abandoned; the real abandonment is not by God of humankind, but that we have abandoned and lost touch with what really matters, and we suffer accordingly.

Additionally, on top of the intrinsic or 'existential' challenges to any faith, there are further challenges coming from science, especially to fundamentalist or anthropomorphic conceptions of God. All of this has lead to a current spiritual crisis probably unprecedented in Western cultures for two thousand years, with widespread evidence of a popular spiritual hunger that is both diffuse and multicentric. The old Christian myths have lost much of their grip, their role as what Reiff called the 'cultural symbolic' — those bedrock assumptions about life that form the most vital cement for a culture. Currently, many are alienated from traditional conceptions of religious faith and practice, finding them in varying degrees rigidly doctrinal and theologically obtuse, socially irresponsible, and philosophically problematic. Many have turned to the sciences, especially the psychological sciences, for answers and help around these old questions of meaning, purpose, and human relatedness. In response, traditional faiths have struggled to make room for insights from psychology and the natural sciences, but often times without even addressing, let alone resolving, serious ontological conflicts between science and traditional Western theologies.

Religion, Attachment Mechanisms, and the Costs and Benefits of Magical Thinking

Bridges are being built at a staggering pace between various psychological disciplines and the neurosciences, and there is a basic assumption now that mind is an emergent property of still poorly defined aspects of central nervous system (CNS) organization and activity. But closely associated splits between faith and scientific knowledge may be even harder to bridge than the old Cartesian separation of mind and matter. I would suggest that any potential deep bridging of faith and science requires a shift away from a concept deeply engrained in both Christianity and Islam, of God as a personage, or as a Being among beings, and ultimately a letting go of the idea that God is just some larger, better and more powerful version of a human figure. The basic resistance in much of Christian imagery and theology to giving up this notion is deeply ironic. There is much irony here in Christianity's strong warning against idolatry. Isn't the notion of God as a Being just another form of idolatry, an anthropomorphism of God? As Tillich put it, religion must embrace the 'ground of being' itself (whatever that may be), rather 'a Being among beings', as all beings are limited by definition.

Put in terms consistent with developmental psychology, our long period of infantile dependence, and the manner in which any concept of value (let alone any concept of ultimate value) is inherently and inextricably bound to images of caregivers makes us ill prepared to give up this anthropomorphism of God. *We are thus powerfully neurologically primed to generate this anthropomorphism of God.* Unfortunately, taken to its literal and concrete extreme in many Western theologies, the mandates of attachment drive us to find a bigger and better image of a person, one that must be somehow hidden behind a cosmic veil This is ironic given Christianity's doctrines; this notion of 'God as Person' is perhaps the ultimate, and most deceptive and seductive, form of 'idolatry'. We commit 'idolatry' by ensconcing in human form (in patriarchal Christianity we have a gender-based notion of male figure no less!) what must be the very core of our creation: the birth of the universe from a Big Bang, the genesis of star systems and planets, the creation of life within planetary biospheres, its evolution, and the eventual phylogenesis and ontogenesis of consciousness — all processes that we understand incompletely. To 'hand over' these immensely powerful creative processes to a human figure, a bigger and better version of our parents, is something that we are deeply primed to do.

There is unfortunately no proof whatsoever that there is a Being 'out there,' hidden behind some cosmic veil, a bigger and better version of us, that will protect us from pain, or death, or the despair of a wasted life. This loss of the magical kingdom of infancy and early childhood is a most difficult bit of reality to swallow, such that in our darkest and most frightened moments, we can easily fall back into it. Indeed, wishes for such protection and comfort are with us throughout our entire lives, and they are intrinsic to the deep structure of attachment in the human brain. These wishes for comfort are certainly not the last word on reality, but they surely are profoundly enmeshed in our responses to every deep emotional challenge that life might throw at us.

Ironically, the very threats to a more infantile faith based in magical thinking provide clues about another, much more durable, type of faith. Perhaps the most consistent shock to a more infantile faith (in a divine being that could alter natural laws to save us from ourselves or protect us from the natural order) is nature's utter consistency in following those laws. And when it seems to break them, we always seem to discover new laws that were invisible to us that explain the "break." Of course, our newer and better scientific principles also have exceptions, exceptions that build until there is enough material to recast theory in a new mould. But it is precisely this strange and yet elusive constancy around us (mirroring other constancies within us) that holds the answer to what lies beyond more magical faiths that require God to be a literal parent, a larger version of our own parents. But how can one bridge science and a less magical faith? A key bridging construct lies within the concept of emergent properties, and the awesome hierarchies of organization in nature that science is exposing, versus older simplistic notion that science reveals just an empty, mechanical universe. Nothing could be further from the truth.

An Ontology for Science: Cold Mechanisms versus Hierarchies of Emergent Properties

There is in fact little support even within the 'hardest' of the hard sciences (physics) for an image of the universe as some vast clockwork machine. Quietly, over the last eighty years, there have been important changes in the 'hard' sciences, away from mechanistic metaphors for natural processes. Today we have a physics of vast interacting fields, with quantum effects that defy common sense on the microcosmic scale, and relativistic effects that defy common sense on the large cosmological scale. In the physics of the microcosm, we have the Heisenberg Uncertainty Principle as a basic principle of the quantum world (that we can never know *both* where a particle is and what direction it is travelling). In the physics of the macrocosm, there are relativistic concepts of time, space and matter that make 'common sense' notions about the more mechanistic 'face' of reality turn to Jell-O. We have phenomena such as matter getting heavier and changing size as it approaches the speed of light, black holes from which even light can not escape, the slowing of time in gravitational fields, and the Big Bang beginning from a nearly infinitesimal point. These ideas make the mechanistic and Euclidean face of our common sense, three-dimensional, every day sensory reality turn to mush. Thus modern physics suggests something quite different from a 'clockwork universe'.

Rather than a myopic view of the cosmos as simply mechanistic, science suggests a vast and interlinked chain of dynamic organization and structure, and a train of 'emergent properties' (including at the apex, consciousness itself). Emergent properties are those that cannot be predicted from any detailed understanding of antecedent levels of organization, and indeed, nature appears to be filled with them, totally frustrating our wishes for an all-encompassing science that would explain all phenomena at all levels of organization. This is now quietly but widely acknowledged as impossible. Instead, virtually every level of organization studied in one science provides the foundation for those processes studied in another science. For example, principles outlined in physics describe the foundations for the science of chemistry. Basic particles underpin matter and energy as we understand them, with some particles forming the building blocks of matter, and some forming the foundations for force. Perhaps underneath (and uniting) those basic particles and forces are very strange things physics has called 'superstrings', structures so bizarre that they might contain extra dimensions, folded into unimaginably tiny spaces.

However, in chemistry, subatomic foundations for matter and energy can be taken for granted, due to the relative permanence of atoms and the relative separation of matter and energy at certain ranges of temperature and pressure in chemical systems. At this chemical level of atomic organization, atoms form molecules and these compounds interact with one another according to certain chemical principles and rules. The science of chemistry in turn describes foundations for the science of biology. Again, these foundations and rules governing the now molecular nature of things are in turn taken for granted in biology due to the relative permanence of molecules at the ranges of temperature and pressure in living systems. At higher biological levels of organization, we are now in the "sciences of complexity", which would include neuroscience. In these sciences, one deals with phenomena that have many interacting variables, and chaos theory is one recent attempt to formally conceptualize such non-linear relationships. Life has to be considered an 'emergent property', from standpoint of any understanding of physics or even chemistry, as no level of understanding *confined to those fields alone* could predict that chemical complexes could become self-organizing and self-replicating. Similarly, no understanding of simple DNA or biochemistry alone could allow one to predict the creation of complex central nervous systems, or the emergence of sentience from those biological structures, and thus consciousness itself would also have to be considered an emergent property. Indeed, from the perspective of the very hot early universe containing nothing beyond quarks and a few other elementary particles, just a second or so past the Big Bang, even complex molecules might be considered an emergent property, as they require electrochemical bonds that could not exist without whole atoms, and the electroweak force hadn't even differentiated into two forces (electromagnetism and the weak force involved in nuclear decay), a differentiation that could not yet exist in such extremely high temperatures and pressures.

Thus, the frustrating complexity of Nature that science reveals contains no ultimate, hard, 'physical stuff', just a seamless web of organizing and structure-building processes, spanning from the micro- to the macrocosm, an infinite web of truly wondrous structures. Science allows us to see downward into the microcosm of the infinitely small, upward into the macrocosm of the infinitely large, and at the intermediate zone of structures in the middle — solar systems, planets, and their ecologies. But science also by its very nature encourages us to look separately at each level of structure, diminishing, even sundering, its living connections to the others. If one looks instead at this much larger framework for nature, one sees not a cold, purposeless set of mechanisms, or an empty machine pointing to a now-lost anthropomorphized Creator, but instead the revelation of this vast web of structure. We sit in this intermediate life-giving zone between the microcosm and macrocosm. In this zone, ultra-complex organizations of protein interact with one another while they are gravitationally held together in a nurturing biosphere. This biosphere, and its vast ecology, would never have come into being without the electromagnetic energy radiated by a gravitationally-driven matter-energy conversion system (the sun), which warms the biosphere just enough (but not too much), thus providing the energy necessary to drive complex metabolic cycles. Within these large ecosystems, rest fundamental drivers in the environment for the evolution of various life forms. Perhaps no scientific discovery is more revealing of a vast interplay in between large scale cosmological processes and smaller scale biological processes than the discovery that the heavier elements (required in complex compounds necessary for life) are produced in the interiors of stars that go supernova. We are literally made from the products of stellar furnaces that exploded billions of years ago, explosions that seeded our section of space with the building blocks for life itself. These ideas place science and religion into uncanny and unfamiliar congruity: we have been created by forces of virtually infinite range, subtlety and power.

A Final Challenge —
Science: Ethically Neutral, or Not?

Perhaps a final obstacle to seeing science and religion as fundamentally compatible in terms of their deepest intents is the appearance of science as ethically neutral or without inherent ethical content or implication. There are even some in the sciences who would explicitly advocate for this to be the case; they suggest that to think otherwise would introduce 'biases' that distort scientific objectivity. While there is great merit in maintaining that scientific inquiry should be done without bias (although of course eliminating all biases is impossible and even undesirable, as one must make theoretical assumptions to do any scientific experiment), I think that the idea that science has no inherent ethical content is nonsense. At the very least, science encourages and values a particular way of asking and exploring questions, and offers a critical discipline — the scientific method — for keeping emotional needs and personal prejudices from contaminating one's understanding of the world and one's self. I would want to underline especially that *it is still the only method that we have for achieving that most elusive and almost self-contradictory goal, a picture of the world that is not simply a projection of our wishes and fears.* Such a picture of the world, not based primarily in wishes, fears, or cultural values, is consensually validated by definition, and potentially duplicated by anyone generating the necessary hypotheses and doing the necessary tests. There is simply no way to overstate how critical this is for human progress, no way to overestimate the importance of our ability to do this. Absent this, and there can only be beliefs grounded in those very same and ancient fears and wishes. Absent science, and we would still believe that the sun spins around the earth, and might still bleed the sick, to purge them of their illness.

Perhaps those who would see science as without ethical content might examine implications of the behavioral sciences for our treatment of each other, and also the science of ecology. Curiously, these are mirror images, in that both ecology and any scientifically constructed ethics of human behavior centrally emphasize notions of complex inter-dependence and reciprocity. Ecology teaches us about the profound interdependence of the fabric of life, and that tears in the fabric, while repairable to a point, threaten not just a few species, but our very existence — if the tear is deep enough. We have no idea to what extent homeostatic mechanisms can adjust to the loss of some of the interlocking pieces that make up ecological cycles and their feedback and feedforward loops. While the planet's living systems can and do repair themselves to some degree, there must be a point beyond which the whole living fabric of fantastically complex checks and balances — of one life form supporting another and a complex environmental matrix — begins to unravel. We may not find that line, the point of inducing a terrible cascade of massive 'homeostatic' collapse, until we have crossed well over it. Once over that line, we may not be able to get back, or stop a destructive unraveling of the fabric of life, as the falling dominoes may continue until the biodiversity of the planet is severely damaged. Thus, *we cannot possibly know how much time we have left* in which to stop the ongoing destruction of the environment, which is progressive and deepening year by year. We are already deeply entrenched in a frighteningly dangerous and uncontrolled experiment to see how much the loss of the ozone layer and the increase in greenhouse gases (from air pollution) will effect the structure of life on the planet. Although it would be impossible for human folly to destroy all of life on the planet — it is far too powerful and resilient for that — it is certainly not beyond our power to impoverish life catastrophically.

If there are any questions remaining about the ethical neutrality of science, one merely needs to look at the behavioral sciences, where despite the cacophony of voices, one hears certain things over and over. Only the most nihilistic behavioral scientist would dare argue that there is not overwhelming empirical evidence that children need to feel loved, valued and nurtured. Children that are neglected and abused often become neglecters and abusers. Neglected children's brains do not even develop optimally: we are biologically damaged by emotional trauma and the absence of emotional nurturing. To whatever extent those things were missing we remember that lack deeply, whether we think so or not, for the rest of our lives. The complexities of an individual's life course, examined in depth, show a profound patterning based on early experience. We repeat and re-enact childhood traumas in relationship to others our whole lives, as we are unconsciously trying to win old battles that we lost long ago. We pass on *both* the love and the trauma of our childhood into the next generation, and it requires the most concerted effort and devotion to the task to tilt the balance just little more positively than what was given to us. The psychological sciences also document that lives can be repaired, that we can improve on what our parents gave us. The repair process stands on whatever intact capacities we have for attachment and hard work, and with help, people can and do make remarkable recoveries from terribly damaging trauma. When we hear such stories, of heroic repair, most of us are moved deeply, because it affirms things we need to believe: love and hard work do make a real difference, and even the worse evils can be conquered by our best efforts. Such moving stories suggest also that the need for connection has been 'built-in' at a deep level of the genetic blueprint for human beings that evolution produced. Indeed, affective neuroscience is now suggesting that primitive abilities for attachment go far back into our phylogenetic ancestors, back perhaps even hundreds of millions of years. Evolution discovered that binding creatures together in social groups and pairs enhanced their survival and the survival of their offspring. To use the older anthropomorphic language that is both scientifically problematic but emotionally compelling, God designed us far better than we could ever hope to. The 'design' is alive with elegance and grace; we have but to learn not to violate it.

For all these reasons, the psychological and behavioral sciences do not support any version of 'ethical relativism'. It is not a coincidence that most scientists, particularly the very best and most gifted, tend to be ethically passionate.

A Common Ground in Religious Symbols, Images, Myths

Perhaps one of the saddest elements in the history of human civilization already alluded to are the chronic and intrinsic abuses embedded in the history of most major religious traditions: the tragic lunacy of people killing one another, rationalized in terms of religious belief systems. Instead of the legacy of countless religious wars, our species badly needs to tolerate a diversity of faith. I personally believe that unless we do a far better job of finding crucial emotional 'common denominators' in various faiths, and embracing them as such, this destructive process, surely writ large in our geopolitical situation currently with Christians, Jews and Muslims squaring off against one another, will only continue, probably with escalating body counts given our enhanced technological abilities to kill one another.

But what are essential common denominators in affective terms in the world's religions? What are the potential emotional commonalities obscured by doctrinal differences we have fought over for millennia? One of my deep disappointments in religious studies as an undergraduate was the widespread failure to examine just this central question. I would suggest a very simple set of common affective denominators, and elegantly simple criteria to evaluate the emotional quality of religious belief systems. If a faith works to deepen reverence for life, sustains connections to others and to Nature, provides support and guidance at times of crisis, binds a culture together, and gives one a framework for 'ultimate concerns' as Tillich said, then that faith offers something of enduring value, regardless of whatever theological or philosophical deficiencies others might find in that faith. By those more emotional (and less theological) standards, the world's religious traditions would have to get mixed reviews, as they have all, at different times to varying degrees, divided us from one another and ourselves, separated us from the Creation, and promoted prejudice, war, poverty and social injustice. This should hardly be surprising, as human faiths are human (and not divine) constructions, full of the human vulnerability to prejudice, distortion and rigidity. We seem uniquely, even perversely, talented as a species at fighting over the religious forms, and thus losing the spiritual and emotional substance.

The great myths of the world's disparate faiths ultimately speak to very similar needs and intents. The best in religious myth and symbol speaks to us of the resilience and richness in human life, and of many old and simple truths. Many faiths speak of these simple universal values: that we must care for our children or else we care for nothing at all; that we hold a place of value in an enduring natural order (however one chooses to construct that order); that there is purpose and meaning to living, but that we must struggle terribly at times to find that meaning; that such meaning and purpose cannot exist without the grounding of our attachments to loved and valued others; that we are part of a great chain of being that goes back into a distant past too far to reconstruct, and forward to a future we can shape but never know; that we must be our brother's and sister's keepers; that we return symbiotically in death to that great unknown from which we were all born.

I have already spoken of the centrality of the many myths of endless cycles of life and death, of dying and rising Gods. I am not using the term 'myth' here to emphasize or imply that such theologies are 'false', but only to highlight their essential commonalties. Embodied in the myth of the dying and rising God is perhaps the most critical component of any real faith: *that there are forces at work in the cosmos and in human life that move in the direction of good and against evil that we can neither fully understand nor harm in any permanent way, even in our worst and most destructive moments.* These organizing forces are beyond harm, and beyond good and evil in that sense, and form a true 'Ground of Being' as Tillich once called it. These organizing forces, silently at work at every level of structure in the universe, precede us. They have always been there, and always will be. They have enabled all that we are, indeed, all that there is, and will continue to do so. Evil then is merely our failure to respect those primary organizing forces at work in human life. Evil does harm to living organizations, to 'God's handiwork', but it cannot harm these organizing processes themselves. This faith in those good things at the heart of life truly does sustain and 'save' us, but not by magic, or by the intervention of an invisible Being, but rather by our ability to use and believe in those organizing processes, in all their innumerable and protean manifestations in the real world of our everyday lives. At the level of human social systems (clearly very high in the natural hierarchy), those organizing processes are reflected in empathy and attachment that human beings generate, and also seen in very old values such as the 'Golden Rule', an effort to outline the pivotal notion of reciprocity, among the most critical of all social rules.

'Masks of God':
An Ontology of Science and Images of the Infinite

The deeper one digs into science and asks about its basic epistemology, the more one is left with a radical ontology: *there is only organizing process,* a continuous hierarchy of emergent properties, and no real duality of mind or matter at all. *This is an image from science identical in its deep structure to basic mystical notions.* As Blake said, "if the doors of perception could be cleansed, we would see things for what they really are ...

infinite ...' In this sense, science can only support a kind of deep Pantheism, never an anthropomorphism. If there is only organizing process, then in some sense, what we have looked for in our Masks of God[6] sits under all things in the cosmos, and all things are illusory, covering this deeper reality. This is a truly bizarre place for science to end up!

We must accept that religious belief systems, like all human constructions, are our limited and flawed efforts to understand that which lies obscured in the infinite depths of the creation. All our codifications in religion attempt to grasp at the infinite, in a real sense, an impossible job. Campbell's wonderful metaphor here points to how we both need, and are potentially misled, by our particular 'Mask of God'. Our finite minds cannot comprehend something that is not finite, and words like 'infinite' in fact have little emotional meaning, so we must construct a myth, what Campbell eloquently calls a 'Mask of God'. It both reveals *and* masks what we are seeking to grasp.

In so far as those myths and images activate feelings of reverence, awe, and connection, they have 'done their job'. In so far as we take them as 'literal absolutes', that our particular images are the right and true ones, then we have just missed the forest for all the trees. In so far as we go even further and become willing to attack others who have *different* images of the infinite as 'heathen' or 'infidel', we have already lost and even violated the very essence of what we are seeking. Surely just such deep irony and recurrent tragedy is writ large in our history as a species. Although such harsh in-group/out-group distinctions, often driven or at least enhanced by religious ideologies, were probably evolutionarily selected to promote individual survival by powerfully enhancing group cohesion, these very processes now threaten our survival as a species.

We must change this. To do so will require many things, but perhaps most important will be appreciation for the emotional common denominators in religious searching and belief, getting past the forms to appreciate, the underlying substance, perhaps for the first time as a species. I see many hopeful signs in many places that the re-birthing of a non-doctrinal spirituality may be aiming in just that direction. Reviewing the endless permutations on 'God concepts' in countless religions yields a disconcerting conclusion for anyone enmeshed in theological debate: *it really doesn't matter what theology we embrace, what matters are the affective currents that our particular theology re-enforces*. In the end, all we have are our fragile connections to life and other living things. The rest is just human story telling, and pointless to fight over, as we have done for millennia.

---

[6]   Campbell (1960-8).

Limitations of Rapprochement Between Science and Non-Magical Faiths

Sadly, this particular vision of rapprochement between science and religion does not support the deep and perennial wish that our individual consciousness be eternal. Notions of an individual disembodied consciousness floating up to heaven seem to be just another example of our deep ability for wishful thinking. Countless jokes about what goes on in Heaven underline these implicit contradictions — do we still fight over the remote control? Our individual and quite limited perceptual consciousnesses seem inexorably tied to our biological identities and brain function, and thus it is very unlikely that they are eternal. Additionally, this train of emergent properties that science offers (instead of a hidden Being) seems quite disturbingly indifferent to an individual life. For that matter, even the entire tapestry of life on Earth could be extinguished with just one small black hole or one large asteroid. In the context of such an event, many of us would no doubt feel that God had (again) abandoned us, when in reality, it would be just "cosmically bad luck", reflecting only the unfolding of processes and events built into the fabric of things. Thus, in this strange yet elusive constancy around us and within us mirrored in both science and religion, there are still no guarantees. I will never see what once appeared in a childhood dream of mine, handwriting on the Moon, or any other miracle that violates natural laws and proves that God truly is listening. There are no such things as miracles that violate the order of things. The real miracles are embedded in the natural order and cannot be separated from it.

Thus, the eternal disappointment of any faith or spirituality that forsakes magical thinking (and only such a faith can *ever* bridge to science) is that as badly as we want and need it, *God does not come and sit on our front porch and talk to us in the King's English*. This image of God thus still disappoints us — or perhaps it is all just part of our continuing challenge to truly grow up. But instead of these desperately sought childlike protections and guarantees, the richness of the Creation cries out to us, summoning us to grasp its deep order at every level of structure, its impenetrable wisdom, its unfathomable, hypnotic depths, and its frightening beauty. What we so keenly search for is mostly veiled to us, and we have to struggle to glimpse at the infinite through a dense veil with our limited vision. God truly does seem to be everywhere, yet nowhere. The radical ontology of science suggests that *God is surely in the dirt under our fingernails, but most certainly not guaranteeing that we get all our wishes.*

The Perennial Problem of Basic Human Evils, now Compounded by Technological Ability

We are tragically good at being tribal, good at being defensive and even paranoid about those other tribes. All this just stems from basic affective states that evolution engendered in us and many other creatures to keep us alive, but simple fear and even rage are *not* the true sources of evil. It is more an *hypertrophy of basic fear and rage systems* from trauma, hurt, and the sense of hopelessness created in the face of the loss of basic emotional supplies we still need unconditionally. All human evils, even the greatest and most horrific, can be traced to those simple sources, and to the group and cultural blindness induced by and reinforcing these states. To the extent that we obscure this, we potentially lose sight of the real causes for much of what we do both individually and as a species that is wrong, even those things that are terribly, horribly wrong. Underneath all of our worst wrongdoing, our worst evils, are those hypertrophied fear and rage systems, and our own unacknowledged and unbearable losses, of safety, comfort, connection, and hope for love. Evil in that sense is the result of 'good not realized', indeed evils emerge from basic human needs being profoundly thwarted, and a 'release' of basic fear and rage systems in the brain. These abilities to violate others while in the throes of our basic emotional defensive systems is complemented, indeed compounded, by our inherent territoriality. We can never have enough territory, or enough supplies, particularly if critical emotional supplies are deeply lacking. The compensatory processes we are capable of here may well drive alienated technological homo sapiens to destroy much of the fabric of life on earth.

To the extent that we commit hurtful acts, grounded in our own losses, injuries and empathic failures, we then give others the grief of losses potentially equally unbearable. This principle applies to both the smallest and largest hurts human can inflict on one another. To break the chain of human wrongdoing, we must bear our own losses, and not pass them through to others. In terms of our recent history, the specter of terrorism, with its roots in admixtures of genocide, wrenching poverty, disenfranchisement, and other aspects of cultural 'shear' and persecution, gives us the clearest illustration of this, our latest horrific object lesson in how the sins of *and* against the father are visited on the son.

Some Final Thoughts on the Mandates and Gifts of Attachment

There are primitive networks in the brain (still poorly mapped) that allow the infant and its parent to look at each other in the eye, and smile, and have powerful empathic resonances with each other, as essential foundations for attachment. I do believe that there is no more powerful force at work in human life than what we create in those simplest of moments. A person's day can be full of those simple moments and empathic affirmations, in large and small ways, or utterly devoid of them. If one thinks critically about human life, and its tragedies and triumphs, one is forced to the conclusion that nothing even comes close in terms of its psychological and biological power, and further, that many powerful events derive their influence from this essential, irreplaceable source of value, including many of our most potent myths and religious symbols. And our enduring need for connection to other sentient creatures comes in a thousand guises, made only the more subtle and complex by our extended cognitive capacities. Yet all our greatest technological creations pale against this simplest and most essential achievement, of mutual affirmation, in simple human attachment. Love, and sometimes especially its wrenching absence, really does make the world go round, with the latter often triggering profoundly destructive processes. Despite the endless seduction of all our other powers, without this simple creation (and re-creation) of human connection, all our other great achievements are as empty and purposeless as can be. Without it, we are utterly alone, our mortality becomes unbearable, and, in the old theological language, we have lost our souls.

All of this suggests that any deep examination of the brain (or of virtually any other scientific field) shows us that we have not been 'abandoned by God' and left alone in a mechanical universe, rather that nature and evolution have carved with exquisite detail a deep biological and social valuing informing every aspect of brain function, even though we may at times do a poor job of interpreting those value signals (emotion), and at times a poor job of nurturing the most vulnerable stage of that process in childhood. Some fortunate few of us seem to live lives as though in full and complete possession of these spiritual truths moment to moment, while others live lives testifying to the staggering destruction we are capable of when we lose our way. Most of us, indeed the vast majority of us, fall somewhere in the middle. If as a species we could become just as good at empathizing with others as we have been at being tribal and defensive, just as good at preserving our planet, its resources, and life forms as we traditionally have been at greedily gobbling up territory and resources, just as good at nurturing our children and those less fortunate than ourselves, who knows what we could become millennia from now. Something truly remarkable and wondrous, realizing all the vast potentials that evolution graced us with. However, there is much reason to worry that continuing down our current path will lead to ecological collapse, global famines, and horrific wars in the context of inevitable nationalistic and regional competition for declining resources, as our destruction of the global environment continues at an accelerating pace. Thus any sensitive observer of our planet at this time must feel torn between deep worry and equally deep hope. It seems very hard to predict how the vastly interactive collection of complex brains on our plant (the human race), dancing to the music of our brain's basic emotional systems and our complex social groups, will move in the coming centuries. Will we turn away from, or over, the precipice in front of us? Although time appears short, I can only hope that a deepening, global consciousness of these basic emotional truths of human life will be enough to save us.

Isabel Clarke

There is a Crack in Everything: That's How the Light Gets In[1]

I have got a lot of time for frogs. Their capacity to remain still and alert by the hour, in an unregarded corner of the iris clump; the powerful and graceful strokes with which they propel themselves unexpectedly to the surface from the depth of the pond; the way in which those same strong limbs can take them leaping across the garden to lurk in the dense foliage and hoover up the slugs. I feel honoured when they grace us with their presence, and the glittering gold and brown eyes meet mine across the pond. Their otherness complements my being, and that of my kind.

I am going to argue that we humans have our amphibious nature — not in the physical sense (lets face it, we are not that good at swimming), but in the medium of our consciousness, and that research into cognitive organisation suggests that this duality is written into our cognitive architecture. We have the capacity to be in our bodies, to respond to inner and outer stimulus and participate in the world of connection. This capacity is normally tempered and modified by the verbal, intellectual, capacity to discriminate and anticipate precisely; to experience ourselves self consciously as individuals. In this way we hop on the land of humanness, but I suggest that sometimes, we are able to leave this accustomed mode of being partially or totally behind and swim in the connected whole, like the angels — and our pre-human ancestors: perhaps even the frogs!

[1]   Cohen (1992).

What do I mean by this swimming mode of being? The idea of step-ping into another world; passing through the veil; the ordinary switch-ing and becoming extraordinary; accounts of such experience are familiar enough in the world of story and legend, as well as in spiritual literature. Here are a few examples. The second two illustrate how this experience is lost as mysteriously as it is attained.

> Looking into the inside [of the wardrobe], she saw several coats hanging up — mostly long fur coats. There was nothing Lucy liked so much as the smell and feel of fur. She immediately stepped into the wardrobe and got in among the coats and rubbed her face against them. ... 'This must be a simply enormous wardrobe!' thought Lucy, going still further in. ... Then she noticed that there was something crunching under her feet. 'I wonder is that more mothballs?' she thought, stooping down to feel it with her hand. But instead of feeling the hard, smooth wood of the floor of the ward-robe, she felt something soft and powdery and extremely cold. 'This is very queer', she said, and went on a step or two further. Next moment she found that what was rubbing against her face and hands was no longer soft fur but something hard and rough and even prickly. 'Why, it is just like branches of trees!' exclaimed Lucy. And then she saw that there was a light ahead of her; not a few inches away where the back of the wardrobe ought to have been, but a long way off. Something cold and soft was falling on her. A moment later she found that she was standing in the middle of a wood at night-time with snow under her feet and snowflakes falling through the air. (C S Lewis, *The Lion, the Witch and the Wardrobe*, 1959, pp. 12–13)

> Jesus took Peter, John and James with him and went up into the hills to pray. And while he was praying the appearance of his face changed and his clothes became dazzling white. Suddenly there were two men talking with him; these were Moses and Elijah, who appeared in glory and spoke of his departure; the destiny he was to fulfill in Jerusalem. Meanwhile Peter and his companions had been in a deep sleep; but when they awoke, they saw his glory and the two men who stood beside him. And as these were moving away from Jesus, Peter said to him, 'Master, how good it is that we are here! Shall we make three shelters, one for you, one for Moses and one for Elijah?'; but he spoke without knowing what he was saying. The words were still on his lips, when there came a cloud which cast a shadow over them; they were afraid as they entered the cloud, and from it came a voice: 'This is my Son, my Chosen; listen to him.' When the voice had spoken, Jesus was seen to be alone. (Luke 9. 28–36, *The Bible*)

In my third example Wolfram von Eschenbach describes how Parzifal finds himself in the grail castle.

> They went up to a great hall. A hundred chandeliers all aglow with candles gave light from above to the household assembled there,

and little candles were burning all around on the walls. A hundred couches he saw there — servants whose duty it was had prepared them — and a hundred quilted coverlets lay on them. The couches, set some distance apart, could seat four knights apiece, and in front of each lay a round carpet ...

Next morning, after the grail ceremony, and Parzifal's failure to ask the Fisher King 'What ails you?' he awakes to find the castle deserted.

Yet before Parzifal the warrior mounted his horse, he ran through many of the rooms, calling out for the people, but not a one did he hear or see.

After leaving the empty castle, Parzifal spends many years seeking to find it again.[2]

This universal phenomenon has given rise to the idea that there is another world; another reality that we humans can access — if we are privileged or enlightened enough — a 'real' reality behind the 'shadow' existence of the everyday (to take the image from Plato's cave, where the everyday is characterised as shadows dancing on the wall of a dark cave lit by a fire, and access to reality lies in stepping outside)[3].

Alternatively, the rationalist thinking of the modern era would label these experiences as illusory, and view the folk tales and religious traditions that treasure them as outdated curiosities.

This chapter tackles this conundrum — and another puzzling and paradoxical phenomenon: the persisting fragility and fallibility of the human, individually and collectively. Progress in science, technology, comfortable living and enlightened ideas (for those in the privileged North) proceeds apace. Yet all this progress does not lead to greater happiness, even for the privileged, and certainly not for humankind in general. Fragmentation of relationships and communities; stress and mental health problems of epic proportions; ruthless exploitation of the environment that mortgages our children's future and threatens the very planet that sustains our life, and wars born of greed and the inability to apportion scarce resources equitably; all these simple facts of our time and many more bear witness to our lamentable inability as a human race to manage any aspect of our life on earth well. Progress in science and sophistication is matched by the abandonment of wisdom — or wisdom's abandonment of us.

Perhaps the juxtaposition of wisdom and science gives a clue to the relationship between these two phenomena. I will argue that the existence of two distinct ways of knowing is the key to understanding this. Furthermore, the central argument of this chapter will be that the crack

---

[2]   Wolfram von Eschenbach (1961), pp. 126, 134.
[3]   Plato (1955), p. 278.

referred to in the title is not a split in reality; not a further dimension or domain, but arises from the split in ourselves; more specifically in our cognitive architecture. In the words of Eluard quoted by David Abram in this volume: 'There is another world. It is this one.' I will first track the clues that point towards this conclusion; I will then expand this point with reference to theories founded in cognitive science, and recent developments in the practice of cognitive therapy (the dominant therapy available within the UK National Health Service [NHS]). I will then draw out some of the implications of this way of looking at things.

A Cognitive Way into the Two Worlds Phenomenon

In order to gain some understanding of what might be going on here, I am going to introduce a model of cognitive functioning. This model is called 'Interacting Cognitive Subsystems' (ICS) and can be found in Teasdale and Barnard's book.[4] Detailed cognitive experimentation suggests that the human mind works by different subsystems passing information from one to another and copying it in the process. In this way, each subsystem has its own memory. Different systems operate with different coding, for instance, verbal, visual, auditory. There are higher order systems that translate these codings, and integrate the information. The crucial feature of this model is that there are not one but two meaning making systems at the apex. The verbally coded *propositional subsystem* gives us the analytically sophisticated individual that our culture has perhaps mistaken for the whole. However, the wealth of sensory information from the outside world, integrated with the body and its arousal system is gathered together by what Teasdale and Barnard call the 'implicational subsystem'. This looks after our relatedness, both with others and with ourselves, and so I shall here use the more descriptive name *relational subsystem*. This is on the lookout for information about threat and value in relation to the self — we are, after all, social primates, and where we stand at any one time in the social hierarchy is crucial for our well being, if not, normally, for our survival. We experience 'where we stand' in the form of our current emotion, be it happy contentment, vague apprehension or seething anger.

We are unaware of this 'crack' between our two main subsystems because they work seamlessly together most of the time, passing information between them, so that we can simultaneously take the emotional temperature and make an accurate estimate in any situation. This starts to break down in states of very high and very low arousal. To be human is to know what it is like to be in a flap, and unable to think clearly — because the body has switched to action mode in response to

[4]   Teasdale and Barnard (1993).

perceived threat, and fine grained thought goes out the window. In our dreams, and on falling asleep, we enter another dimension where logic is totally absent. The application of certain spiritual disciplines, or certain substances, can effect this decoupling between the two subsystems in waking life, so affording a different quality of experience where the sense of individuality becomes distorted or merged into the whole. I have adopted the term 'transliminal' from Thalbourne[5] to describe this state, as it is free of the baggage of other descriptors (mystical, psychotic etc.) Because it implies loss of the ability to get one's bearings in a grounded fashion, the transliminal is not a good state to spend too long in. I have argued elsewhere[6] that the phenomenon of psychosis can be understood as becoming stuck in this state (or as Barnard[7] puts it, when the two subsystems become desynchronised).

The Web of Connection and the Relational Subsystem

The relational subsystem can be recognised as the older part of our makeup, that we share with our non human ancestors. It organises the information we receive through our senses, and our response to this information, in such a way as to promote our safety and well being, but also to regulate our relatedness. First of all, this will entail relatedness to the immediate group within which we find ourselves. Early in life, this is the bond between baby and caregiver, that broadens to take in the wider social group. It is within this immediate social context that we most often experience this 'organisation of our relationships' by the relational subsystem. Let me translate that into something recognisable. You come into a room (e.g. a family gathering, or the meeting of a committee of which you are a member) — perhaps you say something. The room falls silent. Everyone looks a bit shifty and uncomfortable and avoids your gaze. What sensations sweep over your body? A sinking feeling in the pit of the stomach? Tension in the shoulders? Meanwhile the mind (dominated by the propositional subsystem) struggles to keep up and flails around looking for an explanation for this negative social event. The opposite, of course, occurs when you meet up with old friends, family etc. after absence and fall into each others arms in the joy of being together again.

This cameo is intended to illustrate the immediacy of that interconnection between people, and the way in which the emotion, which is the glue in the process, sweeps backwards and forwards between the individuals. Going back to the baby, the question has often been posed:

---

[5]   Thalbourne et al (1997).
[6]   Clarke (2001).
[7]   Barnard (2003).

does it make sense to see the baby in isolation? Winnicott, who has much advanced our understanding in this area, claims that 'there is no such thing as a baby'[8] and that the caregiver/baby dyad is the meaningful unit. Winnicott further argues persuasively that the caregiver (often the mother) cannot be seen in isolation. She/he needs the support of a wider system, starting with the partner and extending out to the family group and the society, to hold her (him) in the vital task of launching a new person in the world. I would suggest that this experience of being held in a web of connectedness is not confined to the first year of life, but is an unremarked substrate to our very existence. The maturation and passage myths of our society and our education system encourage us to see ourselves as lonely individuals forging our way in the world; creating ourselves and our important relationships out of nothing. I would suggest that this is a dangerously unbalanced viewpoint. It makes sense in terms of one half, and only one half of our nature. That half is governed by the propositional subsystem, which, with the success of its logical prowess through the edifice of science, has persuaded us that we can do anything on our own. The state of the world around us tells us otherwise. Acknowledging the importance of the relational subsystem, adds another dimension to our understanding of ourselves as humans.

This dimension comprises our experience of being and knowing through the relational subsystem. This side of our nature is porous to other beings: studies in group process[9], and the therapeutic concept of transference illustrate the subtle blending of people in relationship. I am going to argue that this web of interpenetrating relationship extends far beyond our immediate human circle. As well as waves extending outwards to other humans on the planet, there is our relationship with our ancestors and those who come after us. I will expand on the subject of our relationship with the non-human creatures with whom we share our planet and our relationship with the earth. Less definable, but powerful none the less, is the sense that we humans have always had of relationship with powerful but unseen forces. This experience is unverifiable in scientific terms, but undeniable in its persistence. Angels are fashionable at the moment. Demons, spirits, devas, fairies etc. have abounded in every culture. The sense of relationship with God, Goddess, or the ultimate also displays remarkable persistence in the face of absence of scientific proof. I view this as the ultimate transliminal encounter — with that which is both furthest and deepest, but, because it is way beyond the limited grasp of the propositional subsystem, remains as mystery — inherently unknowable in the propositional sense.

---

[8]   Winnicott (1987), p. 88.
[9]   Dallal (1998).

Relationship with the animal kingdom has traditionally shared this transliminal character. Pet owners would doubtless agree with this. Our tribal ancestors were well aware of their interdependence with the animal kingdom. The utilitarian aspect of use of their products was often tempered by the respect of the hunter for its prey, expressed through ritual, and the intimacy with domesticated animals such as cattle that will have come through shared living quarters, (while remaining aware of the often brutal treatment of animals in traditional peasant cultures). The spiritual or archetypal significance of animals continues to feature in the dreams of modern man.[10] Our tribal ancestors recognised their connectivity with the animals as a route to the power of the transliminal, and pantheistic gods are often either wholly or partially animal.

The relationship between humans and the land has also played a huge role in our development, and in particular, our spiritual development. David Abram has much to say on this.[11] I will merely add an observation on the mapping between contrasting types of landscape and the different experience of the propositional and relational ways of being in different societies. In the Middle Eastern cradle of many of our religious traditions, the desert is the place of the spirit, and the city and cultivated land, of the everyday. The fairytale inheritance of our culture places the transliminal in the virgin forest, which was gradually colonised in the push Eastwards that took place, roughly between the tenth and eighteenth centuries, across Europe and Russia. The village in a clearing, surrounded by forest populated by wild beasts and marginal and magical beings, was both the inner and outer reality for many hundreds of years, and this connectedness persists in the transliminal part of our psyche. Devotion to landscape and the natural world survives in our denatured culture, for example in our leisure pursuits of walking and travelling. This often unacknowledged, but I would argue vital relationship, becomes one of pain as we experience helplessness and anger at the despoilation of the earth in service of economic 'progress'.

As our being is interpenetrated by these circles of relationship, so we are defined and created by them. Our knowing of ourselves is bound up with them. Anthony Ryle, and the other theorists of Cognitive Analytic Therapy have developed this idea in a practical way, in the notion of 'reciprocal roles',[12] which is in turn based on the object relations theory idea of the internalisation of important relationships as a part of development. According to this perspective, which works well as the basis

---

[10] Jung has much to say on this subject, see for instance, Jung (1956), p. 327.
[11] See his chapter below.
[12] Ryle (1995).

for facilitating change within therapy, these internalised reciprocal roles continue to dominate our way of operating in the world. Dysfunction within these defining relationships leads to self defeating and stuck patterns of behaviour. Hence, the quality of our relationship with important others, including the earth and the non-human creatures helps to create us. For instance, where we treat the earth, non-human creatures, or vulnerable peoples with exploitation and contempt, this eats away at our own integrity, and our unavoidable involvement in a society which does this is a continuous sore deep within our being. At the same time, people tend to have a sense of their specialness, even where this is experienced as a loss and a yearning. I would connect this with the underlying relationship with God/Goddess/ultimate. Working therapeutically with people who have been exposed to desperately abusing environments from an early age, there is normally a sense of 'I am worth more than this', which the therapist can work with to facilitate healthy growth and healing. We are created at the same time as we create.

Understanding Spirituality

This model says quite a lot about human beings: that they are inherently unstable taken in isolation; that they are continually in flux and subject to the moral quality of their relationships, for instance. It also offers a way of understanding spirituality which makes it integral to the experience of being human. To return to the subsystems, I hypothesize that we encounter a 'spiritual' quality of experience when the relational subsystem is in the ascendant, but without the dominance of self focused emotions. This allows a state of being in relation with the whole, whether mediated by, say, an experience of nature, or a more abstract experience of God or the ultimate. This experience is generally received as ecstatic and awe inspiring in the short term, but also has its terrifying and persecutory aspects. The encounter with God or angels as reported in the bible is characteristically met with terror, and where the individual's return to the 'safe' state of communication between the two organising subsystems is delayed or unavailable, the result can be the disorienting confusion of psychosis. The ability to pass both ways across the threshold (or 'limen') determines the difference between a beautiful experience and a nightmare world where there are no boundaries and therefore no safety.

I have attempted here to explain a number of phenomena that have always interested me about people: proneness to breakdown as well as the fascination of the spiritual and the obstinate imperfection of our institutions, political and other. In doing so, I have stuck closely to theories that come from information processing. These processing systems link the various bits of the brain that we know govern different functions, so that it is obvious that the propositional subsystem will utilize circuits from the neocortex, whereas older and deeper brain regions such as the limbic system and amygdala will be more important for the relational subsystem. Because of the complexity of the connections (and because of bias in what I know about), I am not grounding my claims in neuroscience, only cognitive science.

The View from this Vantage Point.

In one way a model like this is just a lot of words. In another, it gives us a different organising framework for our view of the world. It is like climbing the next mountain along. The view is different from here. It can take time for the mist to clear so that we can take in the full implications. I will here just point out a few of the features that now become visible.

This model offers some explanation for the features of the subdominant mode of logic, which is discussed in other chapters.[13] The verbal, propositional processing system deals in discrimination; in 'either-or', whereas the relational looks for connection and the whole picture. Where the two become relatively decoupled, the 'either-or' faculty recedes; boundaries dissolve, and 'both-and' becomes the dominant mode. This opens the way to the characteristically paradoxical communication found in all religious traditions. The paradoxical stories of Jesus (the last shall be first etc.) and of the Zen masters alike are designed to lead us beyond normal logic into this place. However, as human beings we need both ways of knowing to survive. One is not superior to the other, and both are incomplete in some way. The everyday, where the relational knowing is tempered by the filter of the propositional subsystem, is precise but screens out much of the whole picture. Where the two are in desynchrony, the relational is dominant, and can apprehend the whole unfiltered. However, it cannot do much else with it. In the short term, the effect might be one of ecstatic oneness, and deep meaningfulness. For any longer, the loss of boundaries becomes disorienting and terrifying. Thoughts are no longer private — they can be accessed by others and inserted; there is no way of discriminating between safety and danger. Danger could be anywhere and everywhere. Where everything is meaningful, everything can spell threat, and conducting normal life and relationships becomes a near

---

[13] See Bomford's chapter below.

impossibility. This is the experience of psychosis. As William James says of the transliminal, 'seraph and snake abide there side by side'.[14]

The idea that the transliminal way of knowing is an inescapable part of our being can help to explain the persisting hold that the supernatural and the spiritual have over us, supposedly modern, people. The fascination of logic and science meets a rival attraction in the fascination of the illogical; the untidy phenomena that do not fit the dominant story, such as telepathy and precognition; the world of dreams and myths - the archetypes as identified by Jung — a notable explorer of this territory, and the sacred and the religious — which defies its obituary writers with its persistent vigour. I have always been interested in the places where these two realms of discourse meet. One aspect of this is the interface between the material and the transliminal. This is complex — a part of the landscape, referred to above, that remains shrouded in mist. On the one hand, it is the relational subsystem that connects with the body and the senses; the propositional that is disembodied and wordy. On the other, the traditional dichotomy has been drawn between matter and spirit. People totally caught up in the transliminal, whether through sublime musing or psychosis, are notoriously ill equipped to manage the material world. However, most of our deepest transliminal and relational experiences, sex being an obvious example, are bound up with our physical natures. The meeting point of the magical (transliminal) with the physical in the form of miracles has always fascinated people. Death and whatever lies beyond is the ultimate example of this boundary. Perhaps the fundamental difficulty of fitting these two areas into one coherant framework suggests that we need to accept 'the crack'; the blind spot on the human windscreen.

Accepting this sort of thing does not come easy to us humans. The intrinsic imperialism of the logical human mind ensures that this faculty assumes command; when it does not dismiss the irrational elements of human experience out of hand, it applies its systematising expertise to the problem. The results can be seen, for example, in the various creeds produced by the Christian church, in an effort to encapsulate a deeply paradoxical story in logical propostions.

This duality of processing is not only evident in obviously 'spiritual' areas such as religion. It is much more ubiquitous than that. The human mind has become accustomed to translate constantly between the two ways of knowing, and both language and the material world become entangled in this process. We are so accustomed to live with interlaced layers of imagination and materially grounded reality that we think nothing of it. For instance, consider the place of myth and story in our

---

[14] James (1902), p. 419.

lives. When we watch a film, or read a novel, we know that what we are hearing is not 'true' — yet it would not hold our attention if it did not grab us at an emotional level; that is, to engage our body's physical arousal mechanism, which is designed for our material survival. To do this there needs to be enough parallels to our own predicament for us to make the imaginative leap, but our minds are clearly designed for this translation. It is well recognised that small children are particularly gripped by stories. As we develop, it is the verbal, propositional side that takes root and takes over. The relational subsystem, with its way of knowing, is the substrate, and its early dominance in the child's world, and gradual eclipse can be observed during the process of development. This is to be observed, not deplored — it is essential to the human condition. However, it means that the world of fairies and father Christmas is that much more immediate to the small child, but they are at the same time more at home with the co-existence of the two incompatible worlds, as they are also surrounded by the discourse of cold logic.

Story and myth are much more than mere diversion. On the transliminal as opposed to the scientific side of the transmission of knowledge, myth and legend have been the prime medium. Stories such as the legends of the ancient gods, and folk tales, take the central themes of all human existence — the vagaries of relationships and rivalries; the challenge of growing up and surviving and making your mark in the world; the transliminal dimension gives these commonplace events a numinous gloss, helped by the eruption of the supernatural and magical into the narrative at certain points — not too often, or the identification will break down. The old gods and their myths might have lost their power, but we still queue up for the latest Harry Potter or Phillip Pullman, and our dreams and our headlines are dominated by the characters and storylines from soap operas. Indeed, the dominance of television and globalised media in contemporary cultural life concentrates huge power over this facet of our collective psyche in these means of communication.

This is enormously significant, as our capacity to create and consume myth and story is not just about entertainment. I would argue that it is about the creation of the self, and of the society. For instance, take the concept of self worth, of self esteem. This is much studied in psychology; it is found to be a crucial variable in the understanding of most mental health problems. If we are aware of ourselves as essentially jumped up primates,[15] our self esteem is an index of where we find ourselves in the primate hierarchy. This is about relationship: how we stand in our relationship with our fellow primates, but also how we

[15]   For instance, like Gilbert (1992).

relate to ourselves — and that is often based on the quality of relationship we had with important figures like early caregivers (i.e. Mum and Dad). All this is embedded in the relational subsystem - which carries with it the sense of specialness and superimportance (along with its flip side of devastation and utter worthlessness). People translate all this effortlessly into the material sphere. Since earliest times, humans have adorned themselves with precious metals and jewels — which express in material form that specialness and enduringness that they seek in the shifting sands of human emotional experience. Today, salary and material possessions, over and above what is needed for survival, meet the same need. In a society which fragments rather than nurtures relationship, the need becomes insatiable, and is transmuted into a material greed which is exploited by those seeking power, and is destroying the earth.

To return to the two ways of knowing, the myth or story is governed by the relational, the transliminal way of experiencing. The story by which we live organises meaning. This is in the realm of ideas, but has powerful effects in the physical world as outlined above. It is ideologies that create wars; that raise up, cement and destroy societies. Leaders through the ages have recognised this, and have sought control of the ideology. The Roman Empire recognised Christianity as a subversive ideology and sought to crush it, until Constantine decided it was better to join them, and effected his successful take-over bid (thus bidding goodbye to Jesus, the anti-establishment, spiritual revolutionary). The religious wars, from the crusades against heretics to the struggles between protestants and catholics that gripped Europe during the sixteenth and seventeenth centuries were about political power and state hegemony, but recognising that this could only be achieved through control of the religion/ideology. Our times provide plenty of evidence of this, whether in the form of fascism and communism, Muslim fundamentalism, or the Christian Right in America.

It is easy to feel helpless in the face of these vast forces, but the very volatility of the transliminal substrate on which they are founded gives room for hope to those of us who are in opposition to the dominant story. Building cities, roads and armament factories takes time and resources. Spreading an idea; telling a story; starting a movement does not. There are no guarantees whose story will prevail. The internet which greatly facilitates this process is also notorious for spreading pornography. The transliminal is morally neutral, but the source of power and of change. If our volatile, relational, minds can translate the psychological into the material, we can, if we choose as conscious thinking beings, translate them back - by opening ourselves to relationship with the wider world and the wonder of it, and by seeking our worth and wellbeing in that relationship, and in joyful and loving relationship with our fellow humans. That sound relationship could extend outwards to care for the non human creation and the earth on which we depend. These things could happen. Through the medium of the transliminal things can happen fast. Possibly even fast enough to save our planet for those future generations, with whom we are even now in relationship.

Physics, Logic and The Pluralistic Universe

The majority of authors here have argued for a move to a recognition of ways of knowing that are alternative to any sort of hierarchical model, a conception in which different ways stand alongside each other, rather than having each one superseding the previous ones. In this section we address the question, what sort of universe are we living in, if this is the case?

Many writers have reacted in different ways to the idea of a plurality of spiritual and religious traditions, with a consequent plurality of ways of thinking about mystical experience. It is a view that poses a challenge to our logical construction of the world. Science, in particular, is used to the idea of a hierarchical sequence of steadily more inclusive theories, each one containing the previous as an approximation or special case, with the whole system conforming to a consistent classical logic. However, the co-existence of different ways of knowing that appear, on conventional (hierarchical) ways of thinking, to be inconsistent, suggests that we are somehow suspending the normal laws of contradiction. Rather that seeing the different views as exclusive alternatives, either of which might hold but not both, we are being enjoined to consider that *both* one alternative *and* the other are in some way valid. The phrase 'both/and thinking' has gained currency as a loose way of characterising this. Such a move cuts across the whole tenor of Western (and much non-Western) philosophy, which has been motivated by the quest for a method leading to comprehensive and exclusive truth. So how are we to judge the truth of different and apparently incompatible ways of knowing? What, indeed is meant by 'truth' in this situation?

The idea that truth is absolute has dominated philosophy in the West until recently.[1] This has been the case both for realists (and quasi-realists),[2] associated with the name of Descartes, for whom truth lies in correspondence with a given external reality; and for idealists (associated with the name of Kant) for whom truth lies in the *a priori* preconditions of our own thinking — but for both truth is absolute, because both assume a sharp dichotomy between the internal and the external world, and both rely on classical logical processes. The idea of alternative ways of knowing based on alternative ways of thinking entirely undermines this conception of exclusive truth.

One response to this undermining of classical truth has been the post-modern project of grounding knowing in particular cultural and political contexts.[3] This has been a vital contribution, but there has remained a gap to be filled. While many writers[4] have affirmed a distinction between post-modernism and pure relativism (the total absence of any concept of truth), the precise logical relation between the two and their relation to a belief in absolute reality has remained elusive.

Two developments have clarified this enormously. One is the work by Jorge Ferrer[5] showing that both contextualism (in its post-modern sense) and absolute realism depend on the basic splitting of the world into 'subjective' and 'objective'. Thus it is this split which must be transcended if we are to understand knowing in a non-hierarchical way. His contribution in this volume shows how we can replace this Cartesian/Kantian split by an understanding of ourselves and the world in terms of participation. In this picture, drawn from a strand of philosophy that started with the phenomenology of Merleau-Ponty and Heidegger, a spiritual occurrence is not a private experience of the world, but the enactment of a participatory event that can involve the creative power of all dimensions of the person: body, vital energy, heart, mind, and soul.

A further essential part of Jorge Ferrer's chapter is his argument, supported by detailed textual references, that the knowings of respected spiritual traditions are in fact not compatible with any sort of hierarchi-

---

[1] For a fuller account of this issue in the context of the ICS model of cognition, see Clarke (2001).

[2] Many scientists, such as John Polkinghorne, recognise that we cannot justify an assertion that the world actually is what science says it is, but would argue that the world behaves consistently *as if* that were the case, which is sufficient to support a classical notion of truth.

[3] See, in the present context, Clarke (1996).

[4] E.g. Spretnak (1991).

[5] Ferrer (2002).

cal concept of truth, so that the alternative being considered here really is necessary.

The second development, from within science (including psychology), is the emergence of a new sort of logic which recognises the existence of different contexts but brings them all within a single non-classical way of handling the process of logical deduction. In this sense, the logic itself is context-dependent. These logics were formulated independently as quantum logic in physics and as bilogic[6] in analytical psychology. Whereas the rise of consciousness studies as a scientific discipline legitimised studying subjective states alongside objective one, this development allows us to see the richness of alternatives that transcend the subjective/objective split.

The central role of bilogic in the conceptualisation of mystical experience has been described in pioneering work by Rodney Bomfor.[7] In his chapter in this volume he describes its origin in the work of the psychoanalyst Ignacio Matte Blanco, who introduced the concept of two logics in the human mind. Of these, one is the classical logic prevalent in conscious thinking, the other is the logic of the unconscious which is often in contradiction to the first logic. Matte Blanco called the latter symmetric logic. The co-existence of the two logics explains many anomalies in human thinking, particularly when it is influenced by the emotions. In the depth of the unconscious symmetric logic is paramount and the thinking — or absence of it — that results is closely parallel to the writings of some mystical theologians, particularly those in the neo-Platonic tradition. His development of the interplay of these two logics is the 'key to understanding much that is deeply uncongenial to the spirit of scientific modernism'[8] and discloses a psychological foundation for the dynamical polarities discussed by June Boyce-Tillman in the first chapter.

In the following chapter I show how closely related ideas emerge in physics, in particular in the structure of quantum logic. I move on to describe how recent work in mathematics gives a precise logical framework for context dependent logic that unites the bilogic of analytical psychology with modern quantum logic, and I argue that this form of logic is what is needed to describe (in so far as any verbal description is possible) the operations of the relational subsystem, and hence many types of mystical experience. The links that we have established between the relational subsystem, mystical experience and context dependent logic now bring mysticism into a full dialogue with the sci-

---

[6]   Matte Blanco (1975).
[7]   Bomford (1999).
[8]   See below, p. 120.

entific approach to the world. It is no longer permissible to dismiss it as 'irrational', because quantum theory is, on this understanding, also 'irrational'. Mysticism simply occupies a different area of a continuum that has already been opened up by quantum physics.

Science is itself strengthened by this move, since it no longer has to carry alone all the problems of humanity. Science, and more generally the rational analytical approach of which science is the best example, points towards other ways of knowing and enables us to understand how these all fit together.

The final chapter in this section, by Lyn Andrews, presents an alternative conceptualisation of this situation based her own spontaneous mystical experience. The account is notable for its vivid depiction of her struggles to make sense of something totally outside anything previously encountered, the transformation that this made in her life, and the vision of the universe that emerged from this. She argues that mysticism is related to increasing self awareness and subtle changes in consciousness, which together might be partly or wholly responsible for the different ways of knowing, and thus for the paradoxes that have just been described in this section. Her story enables us to enter deeply into her process. Some of its features are in common with those of the accounts in Elam's chapter in the first section, but her approach is distinctive for the way in which it gives her a way of integrating many of the aspects of science and mystical insight into a greater whole, in which paradox can be understood as the creative, integrative nature of reality. She also draws out both the parallels, and the differences, between her own experience and the accounts of Jorge Ferrer and Isabel Clarke in the previous chapters.

Jorge N. Ferrer

Spiritual Knowing

A Participatory Understanding

We live in a world of rich spiritual diversity and innovation. Spiritual traditions offer disparate and often conflicting visions of reality and human nature. To the modern mind, this is profoundly perplexing: How to account for these important differences when most of these traditions are supposedly depicting universal and ultimate truths? In the wake of this predicament, it is both tempting and comforting to embrace universalist or perennialist visions that, in their claim to 'honor all truths', seem to bring order to such apparent religious chaos. Despite their professed inclusivist stance, most of the prevailing universalist visions in the modern West tend to distort the essential message of the various religious traditions, hierarchically favoring certain spiritual paths over others and raising serious obstacles for spiritual dialogue and inquiry.[1]

In this chapter, I suggest that *spirituality emerges from human cocreative participation in an always dynamic and indeterminate spiritual power*. Furthermore, I argue that this participatory understanding not only makes universal hierarchical rankings of spiritual traditions appear misconceived, but also reestablishes our direct connection with the source of our being and expands the range of valid spiritual choices that we as individuals can make. After offering a participatory account of the nature of spiritual knowing, I provide a pluralistic understanding of not only spiritual paths, but also spiritual liberations and spiritual ultimates. Then I briefly explore some of the emancipatory implications of this participatory turn for interreligious relations, the problem of conflicting truth-claims in religion, the validity of spiritual truths, and spiritual liberation. Finally, I offer some reflections on the dialectic

[1]   See Ferrer (2000b, 2002).

between universalism and pluralism in spiritual studies, and present a more relaxed and fertile spiritual universalism that passionately embraces the rich variety of ways in which we can cultivate and embody the sacred in the world.

Before proceeding further, however, I should stress that although I believe that the vision outlined in this chapter is more sensitive to the spiritual evidence and better honors the diversity of ways in which the sense of the sacred can be expressed, by no means do I claim that it conveys the final truth about the Mystery of being in which we creatively participate. In contrast, my main intention is to open avenues to rethink and live spirituality in a different, and I believe more fruitful, light. Likewise, although I believe that this vision is advantageous for both interreligious relations and individual spiritual growth, it should be obvious that its ultimate value is a practical challenge that needs to be appraised by others as they personally engage it and critically decide whether it fosters their spiritual understanding and blossoming. It is in this spirit of offering, invitation, inquiry, and perhaps skillful means that I advance the ideas of this chapter.

The Participatory Nature of Spiritual Knowing

I see spiritual knowing as a participatory activity.[2] In the context of this chapter, the term 'participatory' has three different but equally important meanings. First, participatory alludes to the fact that spiritual knowing is not objective, neutral, or merely cognitive. On the contrary, *spiritual knowing engages us in a participatory, connected, and often passionate activity that can involve not only the opening of the mind, but also of the body, the heart, and the soul.* Although spiritual events may involve only certain dimensions of human nature, all of them can potentially come into play in the act of *participatory knowing,* from somatic transfiguration to the awakening of the heart, from erotic communion to visionary cocreation, and from contemplative knowing to moral insight, to mention only a few. Second, participatory refers to the role that our individual consciousness plays during most spiritual and transpersonal events. This relation is not one of appropriation, possession, or passive representation of knowledge, but of *communion* and *cocreative participation.* Finally, participatory also refers to the fundamental ontological predicament of human beings in relation to spiritual energies and realities. Human beings are — whether they know it or not — always participating in the self-disclosure of spirit. This participatory predicament is not only the ontological foundation of the other forms of participation,

---

[2]   Ferrer (2000a).

but also the epistemic anchor of spiritual knowledge claims and the moral source of responsible action.

Spiritual phenomena involve participatory ways of knowing that are presential, enactive, and transformative:

1. *Spiritual knowing is presential:* Spiritual knowing is knowing by presence or by identity. In other words, in most spiritual events, *knowing occurs by virtue of being*. To be sure, it may be tempting to explain this knowing by saying that 'one knows X by virtue of being X'. However, this account is misleading because it suggests a knowing subject and a known object, the very epistemic categories that many spiritual events so drastically dismantle. In contrast, spiritual knowledge is often lived as the emergence of an embodied presence pregnant with meaning that transforms both self and world. We could say, then, that *subject and object, knowing and being, epistemology and ontology, are brought together in the very act of spiritual knowing*.

2. *Spiritual knowing is enactive:* Following the groundbreaking work of Varela, Thompson, and Rosch,[3] my understanding of spiritual knowing embraces an enactive paradigm of cognition. Spiritual knowing, then, is not a mental representation of pregiven, independent spiritual objects, but an *enaction*, the bringing forth of a world or domain of distinctions cocreated by the different elements involved in the participatory event. Some central elements of spiritual participatory events include individual intentions and dispositions; cultural, religious, and historical horizons; archetypal and subtle energies; and, as we will see, a dynamic and indeterminate spiritual power of inexhaustible creativity.

3. *Spiritual knowing is transformative:* Participatory knowing is transformative at least in the following two senses. First, the participation in a spiritual event brings forth the transformation of self and world. And second, a transformation of self is usually necessary to be able to participate in spiritual knowing, and this knowing, in turn, draws forth the self through its transformative process in order to make possible this participation. Therefore, one needs to be willing to be personally transformed in order to access and fully understand most spiritual phenomena. The epistemological significance of such personal transformation cannot be emphasized enough, especially given that the positivist denial of such a requisite is clearly one of the main obstacles for the epistemic legitimization of spirituality in the modern West.

---

[3]   Varela et al. (1991).

An Ocean with Many Shores

Having outlined my understanding of spiritual knowing, I want now to introduce a participatory vision of human spirituality that, as we will see, discloses a radical plurality not only of spiritual paths, but also of spiritual liberations and spiritual ultimates.

Let us begin our story departing from a classic perennialist account. Perennialism generally postulates a single spiritual ultimate which can be directly known through a transconceptual, and presumably ineffable, metaphysical intuition. This insight, so the story goes, provides us with a direct access to 'things as they really are', that is, the ultimate nature of reality and our innermost identity. Central to this view is the idea that once we lift the manifold veils of cultural distortions, doctrinal beliefs, egoic projections, sense of separate existence, and so forth, the doors of perception are unlocked and the true nature of self and reality is revealed to us in a flashing, liberating insight. From a classic perennialist perspective, every spiritual tradition leads, in practice, to this identical, single vision. Or to use one of the most popular perennialist metaphors, spiritual traditions are like rivers leading to the same ocean.

Although this metaphor is used by perennialists to imply a cross-cultural spiritual ultimate, I would like to suggest an alternative reading. I propose that most traditions do lead to the same ocean, but not the one portrayed on the perennialist canvas. The ocean shared by most traditions does not correspond to a single spiritual referent or to 'things as they really are', but, perhaps more humbly, to *the overcoming of narrow self-centeredness and thus a liberation from corresponding limiting perspectives.*

With perennialism, then, I believe that most genuine spiritual paths involve a gradual transformation from narrow self-centeredness towards a fuller participation in the Mystery of existence. To be sure, this self-centeredness can be variously overcome (e.g., through the compassion-raising insight into the interpenetration of all phenomena in Mahayana Buddhism, the knowledge of Brahman in Advaita Vedanta, the continuous feeling of God's loving presence in Christianity, the cleaving to God in Judaism, or the commitment to visionary service and healing in many forms of shamanism, to name only a few possibilities). In all cases, however, we invariably witness a liberation from self-imposed suffering, an opening of the heart, and a commitment to a compassionate and selfless life. It is in this spirit, I believe, that the Dalai Lama thinks of a common element in religion:

> If we view the world's religions from the widest possible viewpoint, and examine their ultimate goal, we find that all of the major world

religions ... are directed to the achievement of permanent human happiness. They are all directed toward that goal. ... To this end, the different world's religions teach different doctrines which help transform the person. In this regard, all religions are the same, there is no conflict.[4]

For the sake of brevity, and mindful of the limitations of this metaphor, since most traditions identify the liberation from self-centeredness as pivotal for this transformation, I will call this common element the Ocean of Emancipation.

Furthermore, I concur with perennialism in holding that the entry into the Ocean of Emancipation may be accompanied, or followed by, a transconceptual disclosure of reality. Due to the radical interpenetration between cognizing self and cognized world, once the self-concept is deconstructed, the world may reveal itself to us in ways that transcend conceptualization. Nevertheless, and here is where we radically depart from perennialism, I maintain that there is *a multiplicity of transconceptual disclosures of reality*. Perennialists erroneously assume that this transconceptual disclosure of reality must be necessarily One. In other words, perennialists generally believe that plurality emerges from concepts and interpretations, and that the cessation of conceptual proliferation must then result in a single apprehension of 'things as they really are'.

But to enter the Ocean of Emancipation does not inevitably tie us to a particular disclosure of reality, even if this is transconceptual. In contrast, what the mystical evidence suggests is that there are a variety of possible spiritual insights and ultimates (Tao, Brahman, *sunyata*, God, *kaivalyam*, etc.) whose transconceptual qualities, although sometimes overlapping, are irreducible and often incompatible (personal versus impersonal, impermanent versus eternal, dual versus nondual, etc.). The typical perennialist move to account for this conflicting evidence is to assume that these qualities correspond to different interpretations, perspectives, dimensions, or levels of a single ultimate reality. As I explain elsewhere[5], however, this move not only is unfounded and problematic, but also covertly posits a pregiven spiritual ultimate that is then hierarchically situated over other spiritual goals. A more fertile way to approach the diversity of spiritual claims is, I believe, to hold that *the various traditions lead to the enactment of different spiritual ultimates and/or transconceptual disclosures of reality*. Although these spiritual ultimates may apparently share some qualities (e.g., nonduality in *sunyata* and *Brahmajñana*), they constitute independent religious aims whose

---

[4]   Dalai Lama (1988), p. 12.
[5]   Ferrer (2000b).

conflation may prove to be a serious mistake. In terms of our metaphor, we could say, then, that *the Ocean of Emancipation has many shores*.

The idea of different spiritual "shores" receives support from one of the few rigorous cross-cultural comparative studies of meditative paths. After his detailed analysis of Patañjali's *Yogasutras*, Buddhaghosa's *Visudhimagga*, and the Tibetan *Mahamudra*, D.P. Brown[6] points out that:

> The conclusions set forth here are nearly the opposite of that of the stereotyped notion of the perennial philosophy according to which many spiritual paths are said to lead to the same end. According to the careful comparison of the traditions we have to conclude the following: there is only one path, but it has several outcomes. There are several kinds of enlightenment, although all free awareness from psychological structure and alleviate suffering.

Whereas Brown, Wilber, and other transpersonalists have rightly identified certain parallels across contemplative paths, contextualists scholars of mysticism have correctly emphasized that the enaction of different spiritual insights and ultimates requires specific mystical teachings, trainings, and practices.[7] Or put in traditional terms, particular 'rafts' are needed to arrive at particular spiritual 'shores': If you want to reach the shore of *nirvana*, you need the raft of the Buddhist *dharma*, not the one provided by Christian praxis. And if you want to realize knowledge of Brahman (*Brahmajñana*), you need to follow the Advaitin path of Vedic study and meditation, and not the practice of Tantric Buddhism, devotional Sufi dance, or psychedelic shamanism. And so forth. In this account, the Dalai Lama[8] is straightforward:

> Liberation in which 'a mind that understands the sphere of reality annihilates all defilements in the sphere of reality' is a state that only Buddhists can accomplish. This kind of *moksa* or *nirvana* is only explained in the Buddhist scriptures, and is achieved only through Buddhist practice. (p. 23)

What is more, different liberated awarenesses and spiritual ultimates can be encountered not only among different religious traditions, but also within a single tradition itself. Listen once again to the Dalai Lama

> *Questioner:* So, if one is a follower of Vedanta, and one reaches the state of *satcitananda*, would this not be considered ultimate liberation?
>
> *His Holiness:* Again, it depends upon how you interpret the words, 'ultimate liberation.' The moksa which is described in the Buddhist religion is achieved only through the practice of emptiness. And this kind of nirvana or liberation, as I have defined it above, cannot be achieved even by Svatantrika Madhyamikas, by

[6]   Brown (1986), pp. 266–7.
[7]   e.g., Fenton (1995); Hollenback (1996); Katz (1978).
[8]   Dalai Lama (1988).

Cittamatras, Sautrantikas or Vaibhasikas. The follower of these schools, *though Buddhists,* do not understand the actual doctrine of emptiness. Because they cannot realize emptiness, or reality, they cannot accomplish the kind of liberation I defined previously. (pp. 23–4)

What the Dalai Lama is suggesting here is that the various spiritual traditions and schools cultivate and achieve different contemplative goals. He is adamant in stressing that adherents to other religions, and even to other Buddhist schools, cannot attain the type of spiritual liberation cultivated by his own. There are alternative understandings and awarenesses of emptiness even among the various Buddhist schools: From the Theravadin *pugdala-sunyata* (emptiness of the person; existence of the aggregates) to the Mahayana *dharma-sunyata* (emptiness of the person and the aggregates) and the Madhyamika *sunyata-sunyata* (emptiness of emptiness). And from Dogen's Buddha-Nature = Impermanence to Nagarjuna's *sunyata* = *pratitya-samutpada* or to Yogachara's, Dzogchen's, and Hua-Yen's more essentialist understandings in terms of Pure Mind, Luminous Presence, or Buddhahood (*Tathagatagarbha*). To lump together these different awarenesses into one single spiritual liberation or referent reachable by all traditions may be profoundly distorting. Each spiritual shore is independent and needs to be reached by its appropriate raft.

From Participatory Knowing to Spiritual Cocreation

Although the metaphor of an ocean with many shores is helpful to illustrate the variety of spiritual ultimates, it is ultimately inadequate to convey the participatory and enactive nature of spiritual knowing advanced in this chapter. As with all geographical metaphors, one can easily get the mistaken impression that these shores are pregiven, somehow waiting out there to be reached or discovered. This view, of course, would automatically catapult us back to a kind of perspectival perennialism, which accounts for the diversity of religious goals in terms of different perspectives or dimensions of the same pregiven Ground of Being.[9]

The participatory vision should not then be confused with the view that mystics of the various kinds and traditions simply access different dimensions or perspectives of a ready-made single ultimate reality. This view merely admits that this pregiven spiritual referent can be approached from different vantage points. In contrast, the view I am advancing here is that *no pregiven ultimate reality exists, and that different*

---

[9]   Ferrer (2000b).

spiritual ultimates can be enacted through intentional or spontaneous cocreative participation in an indeterminate spiritual power or Mystery.

To be sure, once enacted, spiritual shores become more easily accessible and, in a way, 'given' to some extent for individual consciousness to participate in. Once we enter the Ocean of Emancipation, spiritual forms which have been enacted so far are more readily available and tend more naturally to emerge (from mudras to visionary landscapes, from liberating insights to ecstatic types of consciousness, etc.). But the fact that enacted shores become more available does not mean that they are predetermined, limited in number, or that no new shores can be enacted through intentional and cocreative participation. Like trails cleared in a dense forest, spiritual pathways traveled by others can be more easily crossed, but this does not mean that we cannot open new trails and encounter new wonders (and new pitfalls) in the always inexhaustible Mystery of being.

It is fundamental to distinguish clearly our position not only from perspectival perennialism but also from spiritual relativism and anarchy. The threat of spiritual anarchy is short-circuited by the fact that there are certain transcendental constraints upon the nature of spiritually enacted realities. In other words, there is a spiritual power or Mystery out of which everything arises which, although indeterminate, does impose restrictions on human visionary participation. As Varela, Thompson, and Rosch[10] suggest in relation to evolution, the key move 'is to switch from a prescriptive logic to a proscriptive one, that is, from the idea that what is not allowed is forbidden to the idea that what is not forbidden is allowed'. In our context, we could say that although there are restrictions that invalidate certain enactions, within these parameters an indefinite number of them may be feasible.

A central task for spiritual inquirers and participants in the interreligious dialogue, then, is the identification of these parameters or restrictive conditions for the enaction of valid spiritual realities. If I ventured to speculate, I would suggest that the nature of these parameters may have to do not so much with the specific contents of visionary worlds, but with the moral values emerging from them, for example, the saintly virtues in Christianity, the perfections (*paramitas*) in Buddhism, and so forth. In this regard, it is noteworthy that, although there are important areas of tension, religions have usually been able to find more common ground in their ethical prescriptions than in doctrinal or metaphysical issues.[11] In any event, the regulative role of such parameters not only frees us from falling into spiritual anarchy, but also, as we will see in the

---

[10] Varela et al. (1991), p. 195.
[11] Küng (1991); Küng and Kuschel (1993).

next section, paves the way for making qualitative distinctions among spiritual insights and traditions.

Admittedly, to postulate that human intentionality and creativity may influence or even effect the nature of the Divine — understood here as the source of being — may sound somewhat heretical, arrogant, or even inflated. This is a valid concern, but I should add that it stems from a conventional view of the Divine as an isolated and independent entity disconnected from human agency, and that it becomes superfluous in the context of a participatory cosmology: Whenever we understand the relationship between the divine and the human as reciprocal and interconnected, we can, humbly but resolutely, reclaim our creative spiritual role in the divine self-disclosure.[12]

The idea of a reciprocal relationship between the human and the divine finds precedents in the world mystical literature. Perhaps its most compelling articulation can be found in the writings of ancient Jewish and Kabbalistic theurgical mystics. For the theurgic mystic, human religious practices have a profound impact not only in the outer manifestation of the divine, but also in its very inner dynamics and structure. Through the performance of the commandments (*mizvot*), the cleaving to God (*devekut*), and other mystical techniques, the theurgic mystic conditions Divine activities such as the restoration of the sphere of the *sefirots*, the unification and augmentation of God's powers, and even the transformation of God's own indwelling. As M. Idel[13] puts it, the theurgic mystic 'becomes a cooperator not only in the maintenance of the universe but also in the maintenance or even formation of some aspects of the Deity'.

Furthermore, as both L. Dupré[14] and B. McGinn[15] observe, this understanding is not absent in Christian mysticism. In the so-called affective mystics (Richard of Saint Victor, Teresa of Avila, Jan van Ruusbroec, etc.), for example, we find the idea that the love for God substantially affects divine self-expression and can even transform God himself. In his discussion of Ruusbroec's mysticism, Dupré points out:

> In this blissful union the soul comes to share the dynamics of God's inner life, a life not only of rest and darkness but also of creative activity and light. ... The contemplative accompanies God's own move from hiddenness to manifestation within the identity of God's own life. (p. 17)

And he adds:

---

[12]   cf. Heron (1998).
[13]   Idel (1988), p. 181.
[14]   Dupré (1996).
[15]   McGinn (1996c).

> By its dynamic quality the mystical experience surpasses the mere
> awareness of an already present, ontological union. The process of
> loving devotion *realizes* what existed only as potential in the initial
> stage, thus creating a *new* ontological reality. (p. 20)

Although space does not allow me to document this claim here, I
believe that the idea of a spiritual cocreation — 'one that many have
assumed but few have dared to express'[16] — is also present in devotional Sufism, as well as in many Indian traditions such as Shaivism and
Buddhism. In any event, my intention here is not to suggest the univer-
sality of this notion (which clearly is not the case), but merely to show
that it has been maintained by a variety of mystics from different times
and traditions.

To recapitulate so far, the common ocean to which most spiritual traditions lead is not a pregiven spiritual ultimate, but the Ocean of Emancipation, a radical overcoming of narrow self-centeredness which can
be accompanied by a variety of transconceptual disclosures of reality.
In other words, the Ocean of Emancipation has many spiritual shores or
independent spiritual ultimates, some of which are enacted by the
world spiritual traditions, and others whose enaction may presently
require a more creative participation. Although there are certain constraints on their nature, the number of feasible enactions of spiritual
worlds and ultimates may be, within these boundaries, virtually limit-
less. In a participatory cosmos, human intentional participation creatively channels and modulates the self-disclosing of Spirit through the
bringing forth of visionary worlds and spiritual realities. Spiritual
inquiry then becomes a journey beyond any pregiven goal, an endless
exploration and disclosure of the inexhaustible possibilities of an
always dynamic and indeterminate Mystery. Krishnamurti notwith-
standing, spiritual truth is perhaps not a pathless land, but a goalless
path.

After the Participatory Turn

A full discussion of the manifold implications of the participatory turn
for transpersonal and spiritual studies lies beyond the scope of this
chapter. However, I would like to mention at least a few of them in relation to the following four basic subjects: (1) the ranking of spiritual traditions; (2) the problem of conflicting truth-claims in religion; (3) the
validity of spiritual truths; and (4) the very idea of spiritual liberation.

---

[16] Dupré (1996), p. 22.

On Ranking Spiritual Traditions

The participatory turn has important ramifications for our understanding of interreligious relations. Most spiritual gradations stem from the postulation of an ultimate referent from which the relative, partial, or lower value of religious systems and insights is assigned. In terms of our metaphor of an ocean with many shores, we could say that, after reaching a previously laid down spiritual shore or enacting a new one, mystics have typically regarded other shores as incomplete, inferior, or simply false. As the history of religions documents, however, there is no agreement whatsoever among mystics about either the nature of this spiritual ultimate or this hierarchy of spiritual insights. This lack of consensus, of course, is not only one of the most puzzling riddles in philosophy of religion, but also an overriding source of debate in contemporary interreligious dialogue. What is even more important, the idea of a universal spiritual ultimate for which traditions compete has profoundly affected how people from different creeds engage one another, and, even today, engenders all types of religious conflicts, quarrels, and even holy wars. Before suggesting a tentative solution to such a conundrum, and in order to grasp its complexity and pervasiveness, I first offer a few cross-cultural examples of spiritual gradations.

Hierarchical gradations of spiritual traditions have been developed in all major religious traditions. As is well known, Christianity often regarded previous pagan religions as incomplete steps towards the final Christian revelation. Likewise, in Islam, the teachings of Jesus and the ancient prophets of Israel are recognized as relatively valid but imperfect versions of the final Truth revealed in the Koran.

The profusion of alternative spiritual gradations in Hinduism is also well known. For example, while Sankara subordinates the belief in a personal, independent God (*Saguna Brahman*) to the nondual monism of Advaita Vedanta, Ramanuja regards the monistic state of becoming Brahman as a stage 'on the way to union with [a personal] God'[17] and claimed that the entire system of Advaita Vedanta was resting on wrong assumptions.[18] But there is more: Udayana, from the Nyaya school, arranged the rest of Hindu systems into a sequence of distorted stages of understanding of the final truth embodied in his 'ultimate Vedanta', which holds the ultimate reality of the 'Lord' (*isvara*).[19] And 'Vijñanabhioksu, the leading representative of the revival of classical Samkhya and Yoga in the sixteenth century, states that other systems

[17]   Zaehner (1960/1994), p. 63.
[18]   Thibaut (1904).
[19]   Halbfass (1991).

are contained in the Yoga of Patañjali and Vyasa just as rivers are preserved and absorbed by the ocean'.[20] As any scholar of Hinduism can
easily realize, these examples could be endlessly multiplied.

In the Buddhist tradition we also find a number of conflicting hierarchies of spiritual insights and schools. As R.E. Buswell and R.M.
Gimello[21] point out,

> Buddhist schools often sought to associate particular stages along
> the marga [the path], usually lower ones, with various of their sec-
> tarian rivals, while holding the higher stages to correspond to their
> own doctrinal positions. … The purpose of such rankings was not
> purely interpretive; it often had an implicit polemic thrust.

We have already seen, in the words of the Dalai Lama, how Tibetan
Buddhism considers the Theravadin and Yogacarin views of emptiness
as preliminary and incomplete. It is important to stress that, for Tibetan
Buddhists, their understanding of emptiness is not merely different but
more refined, accurate, and soteriologically effective. Needless to say,
this is not an opinion shared by representatives of other Buddhist
schools, which consider their doctrines complete in their own right, and
sufficient to elicit the total awakening described by the Buddha. To
mention only one other of the many alternative Buddhist hierarchies,
Kukai, the founder of the Japanese Shingon, offered a very exhaustive
ranking of Confucian, Taoist, and Buddhist systems culminating in his
own school.[22] In Kukai's 'ten abodes for the mind' (*jujushin*), Buswell
and Gimello[23] explain,

> the fourth abiding mind corresponds to the Hinayanists, who recog-
> nize the truth of no-self … whereas the sixth relates to the
> Yogacarins, who generate universal compassion for all. Kukai's
> path then progresses through stages corresponding to the Sanron
> (Madhyamika), Tendai (T'ien-t'ai), and Kegon (Hua-yen) systems,
> culminating in his own Shingon Esoteric school.

Only in the tenth stage, corresponding to the Shingon school, Kukai
considers the Buddhist practitioner fully liberated.

Interestingly enough, contemporary discussions of spiritual gradations strikingly mirror some of these ancient debates. For example,
whereas Ken Wilber[24] tries to persuade us (à la Sankara) of the more
encompassing nature of nonduality when contrasted to dual and theistic traditions, the theo-monistic model establishes (à la Ramanuja) a

[20]   Halbfass (1988), p. 415.
[21]   Buswell and Gimello (1992), p. 20.
[22]   see, e.g., Kasulis (1988).
[23]   Buswell and Gimello (1992), p. 20.
[24]   Wilber (1995).

mystical hierarchy where nondual, impersonal, and monistic experiences are subordinated to dual, personal, and theistic ones:

> It is possible in a theistic teleological framework to account for monistic experiences in terms of the nature of theistic experiences, treating these as necessary and authentic experiences in the mystic theology. But the reverse does not hold true in a monistic framework. In a monistic framework theistic experiences are not regarded as necessary to the monistic ideal.[25]

What at first sight is more perplexing about these rankings is that their advocates, apparently operating with analogous criteria (such as encompassing capacity), reach radically opposite conclusions about the relationship between nondual and theistic spirituality. When examined more closely, however, this should not be too surprising. The criteria proposed are often vague enough that they can be interpreted to favor one's preferred tradition upon the rest. Take, for example, Wilber's guideline that 'a higher level has extra capacities than previous ones'. Obviously, what counts as 'extra capacities' can be, and actually is, differently judged by the various authors and traditions according to their doctrinal commitments (e.g., nonduality versus the personal and relational qualities of the Divine).

After the participatory turn, however, these interreligious rankings can be recognized as parasitic upon the Cartesian-Kantian assumption of a universal and pregiven spiritual ultimate relative to which such judgments can be made. To put it another way, these interreligious judgments make sense *only* if we first presuppose the existence of a single noumenal or pregiven reality behind the multifarious spiritual experiences and doctrines. Whenever we drop this assumption, however, the very idea of ranking traditions according to a paradigmatic standpoint becomes both fallacious and superfluous. Do not misunderstand me. I am not suggesting that spiritual insights and traditions are incommensurable, but merely that it may be seriously misguided to compare them according to any pre-established spiritual hierarchy. The next sections suggest some directions where these comparative grounds can be sought.

The Problem of Conflicting Truth Claims in Religion

Closely related to the ranking of traditions is the so-called problem of conflicting truth-claims in religion. Roughly, this problem refers to the incompatible ultimate claims religious traditions make about the

---

[25] Stoeber (1994), pp. 17–18.

nature of reality, spirituality, and human identity.[26] Since all religions have been imagined to aim at the same spiritual end, the diversity of religious accounts of ultimate reality is not only perplexing, but also conflicting and problematic.

Although with different nuances, the attempts to explain such divergences have typically taken one of the three following routes: dogmatic exclusivism ('my religion is the only true one, the rest are false'), hierarchical inclusivism ('my religion is the most accurate or complete, the rest are lower or partial'), and ecumenical pluralism ('there may be real differences between our religions, but all lead ultimately to the same end'). Alternatively, to put it in different terms, contextualist scholars invoke conceptual frameworks, and perennialists appeal to hierarchical gradations of traditions and/or to esotericist, perspectivist, or structuralist explanations.

After the participatory turn, however, a more satisfactory response to this conundrum naturally emerges. In short, my thesis is that once we give up Cartesian-Kantian assumptions about a pregiven or noumenal spiritual reality common to all traditions, the so-called problem of conflicting truth-claims becomes, for the most part, a pseudoproblem. In other words, the diversity of spiritual claims is a problem *only* when we have previously presupposed that they are referring to a single, ready-made spiritual reality. However, if rather than resulting from the access and visionary representation of a pregiven reality, spiritual knowledge is enacted, then spiritual truths need no longer be conceived as 'conflicting'. Divergent truth-claims are conflicting *only* if they intend to represent a single referent of determined features. As S.M. Heim[27] suggests in relation to the various spiritual fulfilments: "Nirvana and communion with God are contradictory only if we assume that one or the other must be the sole fate for all human beings". But if we see such a spiritual referent as malleable, indeterminate, and open to a multiplicity of disclosures contingent on human creative endeavors, then the reasons for conflict vanish like a mirage. In this light, the threatening snake we saw in the dark basement can now be recognized as a peaceful and connecting rope.

In short, by giving up our dependence on Cartesian-Kantian premises in spiritual hermeneutics, religious traditions are released from their predicament of metaphysical competition and a more constructive and fertile interreligious space is naturally engendered. To break the Cartesian-Kantian spell in spiritual studies, that is, leads to affirming the uniqueness and legitimacy of each tradition in its own right, and

---

[26] see, e.g., Christian (1972); Griffiths (1991); Hick (1974, 1983).
[27] Heim (1995), p. 149.

only from this platform, I believe, can a genuine interreligious dialogue be successfully launched. The diversity of spiritual truths and cosmologies, then, rather than being a source of conflict or even cause for considerate tolerance, can now be reason for wonder and celebration. Wonder in the wake of the inexhaustible creative power of the self-unfolding of being. And celebration in the wake of the recognition of both our participatory role in such unfolding, and the emerging possibilities for mutual respect, enrichment, and cross-fertilization out of the encounter of traditions.

Two points need to be stressed at this point of our discussion. First, in my view, spiritual pluralism does not exist only at a doctrinal level, but at a metaphysical one. Plurality is not merely an exoteric diversion, but fundamentally engrained in the innermost core of each tradition. As R. Panikkar[28] forcefully puts it: 'Pluralism penetrates into the very heart of the ultimate reality'. And second, there is not a necessary, intrinsic, or a priori hierarchical relationship among the various spiritual universes. There is no final, privileged, or more encompassing spiritual viewpoint. There is neither a 'view from nowhere',[29] nor a 'view from everywhere'.[30] No human being can claim access to a God's eye that can judge from above which tradition contains more parcels of a single Truth, not because this Truth is noumenally inaccessible, but because it is intrinsically indeterminate, malleable, and plural.

By way of concluding this section, I should stress that my defense of many viable spiritual paths and goals does not preclude the possibility of equivalent or common elements among them. In other words, although the different mystical traditions enact and disclose different spiritual universes, two or more traditions may share certain elements in their paths and/or goals (e.g., belief in a personal Creator, attention training, ethical guidelines, etc.). In this context, H.M. Vroom's proposal of a 'multicentered view of religion'[31] that conceives traditions as displaying a variety of independent but potentially overlapping focal points should be seriously considered. As I see it, this model not only makes the entire search for a 'common core' simplistic and misconceived, but also avoids the pitfall of strict incommensurability of spiritual traditions, thus paving the way for different forms of comparative scholarship.

---

[28] Panikkar (1984), p. 110.
[29] Nagel (1986).
[30] Smith (1989).
[31] Vroom (1989).

The Validity of Spiritual Truths

Regarding the thorny issue of the validity of spiritual insights, I should say that the criteria stemming from a participatory account of spiritual knowing can no longer be simply dependent on the picture of reality disclosed, but on the kind of transformation of self, community, and world facilitated by their enaction and expression. That is, once we fully accept the creative link between human beings and the real in spiritual knowing, judgments about how accurately spiritual claims correspond to or represent ultimate reality become nearly meaningless. The goal of contemplative systems is not so much to describe, represent, mirror, and know, but to prescribe, enact, embody, and transform. Or to put it in terms of the Buddhist notion of skillful means (*upaya*): 'The chief measure of a teaching's truth or value is its efficacy unto religious ends, rather than any correspondence with facts'.[32] In other words, *the validity of spiritual knowledge does not rest in its accurate matching with any pregiven content, but in the quality of selfless awareness disclosed and expressed in perception, thinking, feeling, and action.*

It cannot be stressed strongly enough that to reject a pregiven spiritual ultimate referent does not prevent us from making qualitative distinctions in spiritual matters. To be sure, like beautiful porcelains made out of amorphous clay, traditions can not be qualitatively ranked according to their accuracy in representing any original template. However, this does not mean that we cannot discriminate between more evocative, skillful, or sophisticated artifacts. Grounds to decide the comparative and relative value of different spiritual truths can be sought, for example, not in a prearranged hierarchy of spiritual insights or by matching spiritual claims against a ready-made spiritual reality, but by assessing their emancipatory power for self and world, both intra- and interreligiously. By the *emancipatory power* of spiritual truths I mean their *capability to free individuals, communities, and cultures from gross and subtle forms of narcissism, egocentrism, and self-centeredness.* In very general terms, then, and to start exploring these potential qualitative distinctions, I believe that we can rightfully ask some of the following questions: How much does the cultivation and embodiment of these truths result in a movement away from self-centeredness? How much do they lead to the emergence of selfless awareness and/or action in the world? How much do they promote the growth and maturation of love and wisdom? To what degree do they deliver the promised fruits? How effective are they in leading their followers to harmony, balance, truthfulness, and justice within themselves, their communities, and towards the world at large? And so forth.

---

[32]   Buswell and Gimello (1992), p. 4.

Apart from their emancipatory power, there is another orientation relevant to the making of qualitative distinctions among spiritual insights. Essentially, I see the project of constructing frameworks to portray a supposedly pregiven reality the hallmark of false knowing, that is, the pretension of a proud mind to represent a ready-made reality without the collaboration of other levels of the person (instinct, body, heart, etc.), which, I believe, are essential for the construction of genuine knowledge. In contrast, I propose that an enaction of reality is more valid when it is not only a mental-spiritual matter, but a multidimensional process that involves all levels of the person. Although space does not allow me to elaborate this point here, I believe that we are in direct contact with an always dynamic and indeterminate Mystery through our most vital energy. When the various levels of the person are cleared out from interferences (e.g., energetic blockages, bodily embedded shame, splits in the heart, pride of the mind, and struggles at all levels), this energy naturally flows and gestates within us, undergoing a process of transformation through our bodies and hearts, ultimately illuminating the mind with a knowing that is both grounded in and coherent with the Mystery. Because of the dynamic nature of the Mystery, as well as our historically and culturally situated condition, this knowing is never final, but always in constant evolution.

This being said, qualitative distinctions can be made among the various enactions by not only judging their emancipatory power, but also discriminating how grounded in or coherent with the Mystery they are. For example, it is likely that, due to a number of historical and cultural variables, most past and present spiritual visions are to some extent the product of dissociated ways of knowing — ways of knowing that emerge predominantly from the mental access to subtle dimensions of transcendent consciousness, but that are ungrounded and disconnected from vital and immanent spiritual sources. This type of spiritual knowledge, although certainly containing important and genuine insights, is both prey to numerous distortions and, at best, a partial understanding that claims to portray the totality. As I experience it, our lived engagement with both transcendent *and* immanent spiritual energies not only renders a priori hierarchical spiritual gradations obsolete, but also provides an orientation to critique more or less dissociated constructions.

Therefore, a sharp distinction needs to be drawn between 'knowledge that is matched with a pregiven reality' and 'knowing that is grounded in, aligned to, or coherent with the Mystery'. As I see it, the former expression inevitably catapults us into objectivist and representational epistemologies in which there can exist, at least in theory, one single most accurate representation. The latter expressions, in contrast, as well as my understanding of truth as attunement to the unfolding of being, emancipate us from these limitations and open us up to a potential multiplicity of visions that can be firmly grounded in, and equally coherent with, the Mystery. This is why there may be a variety of valid ontologies which nonetheless can be equally harmonious with the Mystery and, in the realm of human affairs, manifest through a similar ethics of love, compassion, and commitment to the blooming of life in all its constructive manifestations (human and nonhuman).

Two important qualifications need to be made about these suggested guidelines. The first relates to the fact that, as the Dalai Lama[33] stresses, some spiritual paths and liberations may be more adequate for different psychological and cultural dispositions, but this does not make them universally superior or inferior. The well-known four yogas of Hinduism (reflection, devotion, action, and experimentation) come quickly to mind in this regard, as well as other spiritual typologies that can be found in both Hinduism and other traditions.[34] The second refers to the complex difficulties inherent in any proposal of cross-cultural criteria for religious truth.[35] It should be obvious, for example, that my emphasis on the overcoming of narcissism and self-centeredness, although I believe it central to most spiritual traditions, may not be shared by all.

These and other difficulties make it imperative to stress the very tentative and conjectural status of any cross-cultural criteria for spiritual truth. But there is more. I do not think that any resolution about cross-cultural spiritual criteria can be legitimately attained by scholars, religious leaders, or even mystics on a priori grounds. What I am suggesting is that the search for criteria for cross-cultural religious truth is not a logical, rational, or even spiritual problem to be solved by isolated individuals or traditions, but a *practical task* to be accomplished in the fire of interreligious dialogue and in actual practices and their fruits.

Spiritual Liberation

The thesis of a plurality of spiritual ultimates also has important implications for our understanding of spiritual liberation. Traditionally, spiritual liberation is said to involve two interrelated dimensions: (1) *Soteriological–phenomenological*, or the attainment of human fulfillment, salvation, redemption, enlightenment, or happiness, and (2) *epistemological–ontological*, or the knowledge of 'things as they really are', ultimate reality, or the Divine. Interestingly, according to most traditions, there is a relation of mutual causality and even final identity between

---

[33] Dalai Lama (1988, 1996).
[34] Beena (1990); Smith (1994).
[35] Vroom (1989); Dean (1995).

these two defining dimensions of spiritual emancipation: To know is to be liberated, and if you are free, you know.

It has been my contention that, although most traditions concur in that liberation implies an overcoming of limiting self-centeredness and associated restricted perspectives, this can be cultivated, embodied, and expressed in a variety of independent ways. Likewise, I also advanced the even more radical thesis that this spiritual plurality is not only soteriological or phenomenological, but also epistemological, ontological, and metaphysical. Put simply, there is a multiplicity of spiritual liberations *and* spiritual ultimates. The tasks remain, however, to address the tension between our account and the traditional claim that liberation is equivalent to knowing 'things as they really are', as well as to explore the implications of our viewpoint for spiritual blossoming.

To begin with, I should admit straight off that this tension is a real one, and that the participatory vision will probably not be acceptable to those who firmly believe in the exclusive or privileged truth of their religions. While respecting the many thoughtful and sensitive individuals who maintain exclusivist or inclusivist stances, I see these pretensions as problematic assumptions that not only cannot be consistently maintained in our pluralistic contemporary world, but also frequently lead to a deadlock in the interfaith dialogue.

The many arguments showing the untenability of religious absolutism are well known and need not be repeated here.[36] To the standard ones, I would like to add that religious absolutism is inconsistent with the nature of spiritual liberation as maintained by those religious traditions themselves. Most traditions equate spiritual liberation with boundless freedom. But if we rigidly maintain the exclusive Truth of our tradition, are we not binding ourselves to a particular, limited disclosure of reality? And if we tie our very being to a singular, even if transconceptual, disclosure of reality, then, we can rightfully wonder, how truly boundless is our spiritual freedom? Is this freedom truly boundless or rather a subtle form of spiritual bondage? And if so, is this the promised spiritual freedom we are truly longing for?

As I see it, the apparent tension between the participatory vision and the mystical claims of metaphysical ultimacy can be relaxed by simultaneously holding that (1) all traditions are potentially correct in maintaining that they lead to a direct insight into 'things as they *really* are', and (2) this 'really' does not refer to a Cartesian pregiven reality. Despite our deep-seated dispositions to equate Cartesian objectivity with reality, it is fundamental to realize that to reject the idea of a pregiven world is not to say good-bye to reality, but to pave the way for

---

[36] see, e.g., Panikkar (1984); Wiggins (1996).

encountering it in all its complexity, dynamism, and mystery. From this perspective, the expression 'things as they really are' is misguided only if understood in the context of objectivist and essentialist epistemologies, but not if conceived in terms of participatory enactions of reality free from egocentric distortions. After all, what most mystical traditions offer are not so much descriptions of a pregiven ultimate reality to be confirmed or falsified by experiential evidence, but prescriptions of ways of 'being-*and*-the-world' to be intentionally cultivated and lived.[37] The descriptive claims of the contemplative traditions primarily apply to the deluded or alienated ordinary human predicament, as well as to the various visions of self and world disclosed throughout the unfolding of each soteriological path. But since there are many possible enactions of truer and more liberated self and world, it may be more accurate to talk about them not so much in terms of 'things as they really are', but of 'things as they really *can* be' or even 'things as they really *should* be'.

In any event, it should be clear that when I say that this 'really' refers to an understanding of reality free from the distorting lenses of narcissism and self-centeredness, I am not limiting contemplative claims to their phenomenological dimension. On the contrary, a participatory epistemology can fully explain, in a way that no Cartesian paradigm can, why most traditions consider these two dimensions of liberation (phenomenological and ontological) radically intertwined. *If reality is not merely discovered but enacted through cocreative participation, and if what we bring to our inquiries affects in important ways the disclosure of reality, then the fundamental interrelationship, and even identity, between phenomenology and ontology, between knowledge and liberation, in the spiritual search stops being a conundrum and becomes a natural necessity.* If this is the case, there is no conflict whatsoever for the participatory vision to simultaneously maintain that there exist a plurality of spiritual ultimates and that all of them may disclose 'things as they really are'.

A More Relaxed Spiritual Universalism

In this chapter, I have introduced a participatory spiritual pluralism as a more adequate metaphysical framework than the perennialism typical of most transpersonal works. If I have argued so forcefully for a multiplicity of spiritual liberations and ultimates it is because spiritual pluralism (1) is more consistent with my own participatory understanding of spiritual states of discernment, (2) affirms, supports, and legitimizes the largest number of spiritual perspectives on their own terms, (3) provides a more fertile ground for a constructive and egalitarian interfaith

[37]   Ferrer (1998, 2001).

inquiry and dialogue, and (4) is more generous in terms of recognizing the infinite creativity of Spirit than other meta-perspectives, allowing, impelling, and catalyzing Spirit's creative urges through human embodied participation. Although for these and other reasons my work emphasizes the metaphysical plurality of spiritual worlds, I should stress here that I do not believe that either pluralism or universalism per se are spiritually superior or more evolved. And it is now time to make explicit the kind of spiritual universalism implicit in the participatory vision.

There is a way, I believe, in which we can legitimately talk about a shared spiritual power, one reality, one world, or one truth. On the one hand, the discussion about whether there is one world or a multiplicity of different worlds can be seen as ultimately a semantic one, and metaphysically a pseudoproblem. On the other hand, a shared spiritual ground needs to be presupposed to make interreligious inquiry and dialogue possible and intelligible. After all, traditions do understand each other and frequently developed and transformed themselves through rich and varied interreligious interactions. The strict incommensurability of traditions needs to be rejected on logical, pragmatic, and historical grounds. Thus, it may be possible to talk about a common spiritual dynamism underlying the plurality of religious insights and ultimates. But let us be clear here, this spiritual universalism does not say that the Tao is God, that emptiness (*sunyata*) is structurally equivalent to Brahman, and similar, quite empty I believe, equations. And neither does it suggest the equally problematic possibility that these spiritual ultimates are different cuts, layers, or snapshots of the same pie. As I see it, the *indeterminate nature of Spirit* cannot be adequately depicted through any positive attribute, such as nondual, dual, impersonal, personal, and so forth. This is why, I believe, so many Western and Eastern mystics chose the so-called *via negativa* or apophatic language to talk about the Divine, and why such nonexperiential language was regarded by most traditions as closer to the Divine than any positive statement of its qualities.

The spiritual universalism of the participatory vision, then, does not establish any a priori hierarchy of positive attributes of the divine: Nondual insights are not necessarily higher than dual, nor are dual higher than nondual. Personal enactions are not necessarily higher than impersonal, nor are impersonal higher than personal. And so forth. Since the Mystery is intrinsically indeterminate, spiritual qualitative distinctions cannot be made by matching our insights and conceptualizations with any pregiven features. In contrast, I suggest that qualitative distinctions among spiritual enactions can be made by not only evaluating their emancipatory power for self, relationships, and world, but also discriminating how grounded in or coherent with the Mystery they are. Moreover, because of their unique psychospiritual and arche-

typal dispositions, individuals and cultures may emancipate themselves better through different enactions of the spiritual power, and this
not only paves the way for a more constructive and enriching interreligious dialogue, but also opens up the creative range of valid spiritual
choices potentially available to us as individuals. In sum, *this vision
brings forth a more relaxed and permissive spiritual universalism that passionately embraces (rather than reduces, conflates, or subordinates) the variety of
ways in which the sacred can be cultivated and embodied, without falling into
spiritual anarchy or vulgar relativism.*

The relationship between pluralism and universalism cannot be
characterized consistently in a hierarchical fashion, and even less in
terms of spiritual evolution. While there are 'lower' and 'higher' forms
of both universalism and pluralism (more or less sophisticated, encompassing, explanatory, emancipatory, grounded in the Mystery, etc.), my
sense is that *the dialectic between universalism and pluralism, between the
One and the Many, displays what may well be the deepest dynamics of the
self-disclosing of Spirit.* From the rigid universalism of rational consciousness to the pluralistic relativism of some postmodern approaches, from
perennialist universalism to the emerging spiritual pluralism of the interfaith dialogue, Spirit seems to swing from one to the other pole, from the
One to the Many and from the Many to the One, endlessly striving to
more fully manifest, embody, and embrace love and wisdom in all its
forms. Newer and more embracing universalist and pluralistic visions
will continue to emerge, but the everlasting dialectical movement
between the One and the Many in the self-disclosing of Spirit makes any
abstract or absolute hierarchical arrangement between them misleading.
If I am right about the generative power of the dialectical relationship
between the One and the Many, then to get stuck in or freeze either of the
two poles as the Truth cannot but hinder the natural unfolding of Spirit's
creative urges. This is why, although originally offered in a different context, the following remark by Habermas seems pertinent here: 'The metaphysical priority of unity above plurality and the contextualist priority of
pluralism above unity are secret accomplices'.[38]

To conclude this chapter, I would like to emphasize that it is *only* after
travelling through the tremendously rich, complex, and multifaceted
spiritual waters that we can, I believe, afford to immerse ourselves in
the profound ocean of this more open, fertile, and relaxed universalism;
a universalism that calls to be realized, not so much in isolated individual inner experiences, grandiose visions, or metaphysical intuitions,
but through intimate dialogue and communion with other beings and
the world.

---

[38]   Habermas (1992), pp. 116–7.

Rodney Bomford

Ignacio Matte Blanco and the Logic of God

Introduction

My aim in this chapter is to explain how the idea of *symmetric logic* can enable us to understand the working of the Unconscious, and then extend this understanding, first to the way in which we continually use the classifications of 'same' and 'different', and then to mystical experience, where many aspects of symmetric logic come together.

Symmetric logic is the invention, or discovery, of Ignacio Matte Blanco (1913-95) a psycho-analyst in the Freudian and Kleinian traditions who had also a strong amateur interest in mathematics and in philosophy. Symmetric logic enabled him to discern patterns in the thinking of schizophrenic patients which accounted for assertions and deductions made by them which the ordinary person would regard as nonsensical madness. As he reflected further on this logic, Matte Blanco saw that it provided a wider, more generalised theory that explained the characteristics of unconscious thinking as Freud had formulated them long before.[1] Since feeling and emotion are held by psycho-analytic theory to be the result of unconscious process, symmetric logic sheds light on how emotion affects thinking, on how, therefore, each of us sees the world and responds to events concerning us. Our minds are not purely rational but intimately combine rational, logical thought with all manner of matters of feeling and emotion which are often called simply irrational. Matte Blanco used symmetric logic to show that the irrational has deep patterns of thought and is not therefore, as might otherwise be assumed, wholly random and erratic.

---

[1]   Freud (1915).

The Structure of Symmetric Logic

Matte Blanco's original formulation of the laws of symmetric logic[2] has certain weaknesses and I shall present next my own adaptation of them.[3] First I should explain that by 'logic' I mean a pattern of thought exhibited in the movement from one or more propositions to one or more further propositions, what in classical logic is called a deduction. The 'deductions' made by symmetric logic are often not valid in classical logic and the conclusions do not follow from the premises in such a way as to conform to empirical reality. They are however the conclusions that people actually at times draw, and particularly they are drawn when unconscious process is influential.

Symmetric logic, like classical logic, deals with *propositions* — statements that assert a property of one or more *terms*. Examples are: 'the tree is green' (a proposition with one term), or 'John and James are brothers' (a proposition with two terms). Propositions can be combined with logical connectors such as 'and', 'or' and so on, to give new propositions. There is an important connector in symmetric logic called 'the opposite of' and denoted by '$-$'; so that if 'a' denotes a proposition, '$-a$' denotes the opposite proposition. Finally there are rules for drawing new propositions (deductions) from a given proposition or several given propositions.

The first, and perhaps the most startling, rule of symmetric deduction is the rule of opposition: given a proposition a one can deduce the opposite proposition $-a$. The meaning of this has to be decided in any particular application. As I shall be using it later it means that, for example, the proposition 'The Prime Minister is powerful' implies its opposite, i.e. 'The Prime Minister is powerless'. In classical logic this would be an invalid deduction, but it is of the essence of symmetric logic that it draws deductions invalid classically.

The second, and almost equally surprising, rule of symmetric deduction is the rule of reflection (from which symmetric logic draws its name). Given a proposition about two terms, such as 'John is the father of Mary', we can deduce the proposition obtained by interchanging the terms; in this case 'Mary is the father of John'. In classical logic this rule would only be valid with certain propositional forms, for example, 'John is the sibling of Mary' does imply that 'Mary is the sibling of John'. Again, as the examples just given show, symmetric logic draws deductions that are invalid classically.

There are further rules of deduction dealing with propositions of three or more terms so that all the terms are treated even-handedly. The

---

[2]   Matte Blanco (1975), pp. 35–41.
[3]   Bomford (1998).

only case to be used here is the rule of rotation. This states that a proposition about three terms, A, B and C (in that order) implies also the propositions obtained by reordering the terms as (B,C,A) and (C,A,B). This rule deals 'symmetrically' with the three terms involved, just as the previous rule deals symmetrically with its two terms. An example of the rule of rotation involving relations between a father, his son and his daughter, will be given below.

Logic and the Unconscious

Symmetric logic is irrational by normal standards of rationality. It breaks down distinctions wherever it is applied. Consider for example the proposition, 'Yesterday is before Today'. If this is treated as a two term proposition (about 'yesterday' and 'today') the symmetric deduction is that 'Today is before Yesterday' and it is apparent that temporal distinction is destroyed. Symmetric logic thus used leads to timelessness and timelessness was one of the characteristics that Freud found in the Unconscious. Another characteristic was the non-contradiction of opposites and this too follows from symmetric logic, by the rule of opposition. The writers of mystical theology found both these characteristics in God, who is said to be eternal and in whom all opposites coincide.[4] If theologically the mystical journey is a journey in search of God within the soul, then psychologically it may be seen as an attempt by the conscious mind to enter its unconscious depths, and these two accounts of it are linked by the concept of symmetric logic.

Matte Blanco's theory of the unconscious rests on symmetric logic, but also on the belief that the unconscious does not know individuals, but only sets of individuals. The word 'set' is used here in the mathematical sense to designate a collection of individual items. A set may have any number of individual members within it (and in mathematics there is the empty set which has no members). It is important to note the distinction between an item and the set that consists of just that one item — for one is the item and the other is a set! To illustrate this property of the Unconscious, imagine a person, I will call him John, who had an authoritarian father and who now has trouble in accepting the role of James, his line-manager. John's unconscious knows James as 'the set consisting of James'. This is a sub-set of 'the set of authority figures in John's life'. John's father also belongs to this wider set. According to symmetric logic (the rule of reflection) if a set X is a subset of set Y, then Y is a subset of X, and thus the two sets are identical. Therefore John's

---

[4] The composition of opposites is most clearly expressed by Nicholas of Cusa (1401–64) but it is prefigured in many other writers and notably in the fifth century Pseudo-Dionysius the Areopagite.

unconscious experiences the set consisting of just James as though it were the set consisting of all authority figures in his life, and, by the same reasoning this is identical to the set consisting of just John's father. The consequence is that unconsciously any authority figure is experienced as though he or she were a compound of all previous authority figures, including the original father with whom bad experience began. Furthermore the set of authority figures holds all possible authority figures, including any that might be imagined or dreamed of. If John experiences all these (compounded together) in his relationship with his line-manager then it is not surprising that the relationship is a difficult one. Freud called this phenomenon — of experiencing many in the one — *condensation* and Matte Blanco's theory explains also the similar concept of *displacement*, by which one person is experienced as though he/she were another, emotion thus being displaced from its appropriate object.

In a situation where, for example, every possible authority figure is experienced in one person, Matte Blanco argued that the set of authority figures has become infinite in extent and that this correlates with the degree of intensity of emotion experienced. He called this the *infinitising of feeling* and the unconscious always tends towards such infinities.

At this point it is appropriate to make an important qualification about symmetric logic. It might well seem incredible to think of 'the unconscious' in such concrete terms, almost as though it were a separate person with a very peculiar logic. The qualification is that, though for convenience one may talk of the unconscious making symmetric deductions, a more accurate phrasing would be that such is the way the unconscious appears to act to consciousness. Unconscious process as observed by consciousness appears to follow the patterns of symmetric logic.

Matte Blanco defined[5] five strata of mental life distinguished by the balance of asymmetry and symmetry apparent in each. These strata may be illustrated by a little meditation on the theme 'Man is a Wolf'. Consider somebody who is aware of his aggressive behaviour and impulses. He might quite unemotionally express this awareness by the proposition, 'I have certain wolf-like propensities'. There is no symmetric logic lying behind this statement which therefore belongs in the first stratum.

Rather more emotionally the same person might say 'I am like a wolf'. This assertion goes beyond the purely logical for presumably he is not asserting that he has four legs and a furry coat, only that he is like a wolf in certain respects. There is some symmetric logic in the move from recognising certain aspects as wolf-like to the implicit suggestion that all aspects of himself are wolf-like. This belongs to the second stratum.

With a great deal more emotion the simile in the last example becomes a metaphor: 'I am a wolf'. One who says this may be assumed to feel a real identity between himself and a wolf, even though he is probably quite aware that he is not a wolf in reality. The assertion functions on two levels. On the one hand it means no more than the first stratum equivalent. On the other it expresses this real feeling of identity. This is an example of what Matte Blanco called a 'bi-logical structure'. The same assertion is simultaneously understood in both asymmetric and symmetric ways. This illustrates the third stratum.

In the fourth stratum emotional intensity becomes so great that it results in chaos and the almost complete destruction of rational thought. It would be expressed by somebody running on all fours and howling like a wolf. The metaphoric identity of the third stratum here becomes absolute without any accompanying sense of distinction. This fourth stratum is perhaps most clearly expressed in extreme psychotic disturbance, but it is important to realise that this stratum is within us all. As Shakespeare wrote, 'The lunatic, the lover and the poet are of one imagination all compact'.[6]

The fifth stratum is described by Matte Blanco as a point of total peace for all asymmetries are done away by total symmetric logic. The 'wolf-man' of the fourth stratum is also a 'sheep-man', so to speak. The wolf-man's total aggression is conjoined with total submissiveness. The law of opposition applied to the proposition, 'I am totally aggressive' delivers 'I am totally submissive', and therefore I become totally both aggressive and submissive. The two attributes cancel each other out and there is peace. Matte Blanco describes this stratum as the state of pure being, without any movement or attribute. It appears to be exactly the state that such mystics as the author of the *Cloud of Unknowing*[7] have described.

In the course of describing the five strata I have introduced the concept of the bi-logical structure. In the illustration I have given this is of a single assertion which has simultaneously different significances in the two logics. There are various types of bi-logical structure and Matte Blanco describes another, wherein the two logics are used alternately to deliver a strange conclusion. A schizophrenic patient was bitten by a dog and therefore went to the dentist! The patient saw the dog's teeth as

---

[6] Shakespeare, *A Midsummer Night's Dream* Act V Scene 1.
[7] I have argued at greater length for the parallels between the mystics' God and the psycho-analytic Unconscious in Bomford (1999).

bad in the sense of dangerous. They therefore belonged to the class of
teeth bad in any sense of the word. Being bitten by the dog had made
him 'suffer from bad teeth'. Therefore (the last step is quite logical!) he
went to the dentist.

I began the chapter with a description of symmetric logic as a purely
abstract schema and have then suggested that this schema is found in
unconscious process — or rather that unconscious process appears to
conscious thinking to be using symmetric logic. I now want to point out
another important difference between the two logics and one which has
immense significance. If one is presented with a set of related propositions and works upon them by asymmetric (i.e. classical) logic, the
deductions that can be made are completely predictable. Any two people are bound to reach the same conclusions. Moreover in a sense no
new information can arise in the process of deduction. The conclusions
may be new to the person making the deductions, but they are already
objectively fixed and thus might be said to be contained in the premises.

Symmetric logic is however very different. It does not follow a set
course, as does asymmetry. When a situation is apprehended by the
Unconscious I have argued that it uses symmetric logic on the proposition that might be said to describe the situation. However, the number
of terms in that proposition is not pre-determined. For example suppose a situation that might be described as 'I am in danger'. This might
most naturally be seen as a proposition about 'me' and symmetric logic
would deduce (by the rule of opposition) that also 'I am in a place of
safety'. But it could also be construed as a two-term proposition about
'me' and 'danger' and then symmetric logic would deduce (by the law
of reflection) that 'danger is in me'. This in turn might have more than
one meaning — it could be taken to mean 'I am a dangerous person',
and that suggests interesting thoughts about self-destructiveness.
Alternatively it might be interpreted as 'all the danger in the world
comes from me' and that too can be meaningful.

Where a situation involving several people is concerned, the possibilities multiply. Imagine, for example, a family of father, mother, son and
daughter. All sorts of emotional currents may be at work each more or
less prominently at any given time.

As perceived by each member of the family, each of these 'currents'
may be expressed in a proposition which (intrasubjectively) may be
worked upon by the unconscious using symmetric logic. I will suggest
some, rather simplistic, examples. The mother may feel that in domestic
chores the males do not do their share. Her feeling might be expressed
as 'the males take advantage of the females'. Here there are four people
involved, but the proposition has apparently two terms — 'males' and
'females'. Unconsciously therefore she may also sense the reflected

proposition, 'the females take advantage of the males'. She may be unaware of what this proposition expresses, but the theory is that somewhere in her the corresponding feeling is at work — and it may not be in relation to domestic chores, since the unconscious ignores many details! However, and this is where the freedom of the unconscious is detectable, the same family situation might be grasped by the unconscious as 'the females are unfairly treated'. This is a one-term proposition about 'females'. The law of opposition may therefore be applied and the deduction then is 'the females are favourably treated'. At first sight this might not seem to make much difference, but there is surely a potentially great distinction between the sense of being taken advantage of by somebody specific, and the sense of being simply taken advantage of. The same distinction is apparent in the deductions. This is underlined by a further consideration: the original proposition and the deduction are simultaneously present. The first possibility then is expressed by 'the males are taking advantage of the females' *and* 'the females are taking advantage of the males'. From this an asymmetric step in the argument delivers a new one-term proposition, 'we (i.e. the males and females) are taking advantage of one another'. If symmetric logic is used again, the law of opposition delivers 'we are treating each other fairly', and again a composite proposition arises, 'we together are both fair and unfair'. It may be questionable whether this last very symmetric conclusion has any meaning. I suggest it might be seen as proposing that gender-difference and fairness are ambivalently linked and that this linkage is 'in the air' at a deeply unconscious level. As a celebration of the system, I call propositions expressing such vague, but profound apprehensions 'zero-term propositions'. I intend to convey by that term that there is something present, but not specifically linked to anybody — a predicate without a subject, as it were.

The example above is intended to underline the freedom of the unconscious in choosing the number of terms in a proposition expressing a situation, and of how symmetric logic tends to reduce the number of terms from two to one, perhaps from one to zero. Another instance of this might be imagined as follows: when mother (M) and father (F) first met we might suppose there was a mutual attraction This inter-subjective situation may be intra-subjectively expressed, in the father's case, as 'I am attracted to M' and in the mother's as 'I am attracted to F'. The unconscious of each will perhaps be using the law of reflection — since these appear to be two-term propositions — to deliver belief in each that because he or she is attracted, therefore he or she is found to be attractive by the other. This reciprocation is typical of emotional attraction — though of course there are other emotional currents possible, particularly where the thought, 'I am attractive' — a one-term proposition — is rendered by the law of opposition, 'I am repellent'. However, we may suppose the relationship between M and F prospers. In each the two reciprocating propositions (I am attractive to X, and X is attractive to me) develop in both into 'I love X' *and* 'X loves me'. These may be compounded (asymmetrically) to deliver 'we love each other'. This is now a one-term proposition and the law of opposition may lead to 'we hate each other', expressing the familiar thought that love and hate are often found commingled. Compounding again, the situation may be represented as 'we are loving and hating together'. At a yet greater depth, I suggest tentatively that a zero-term proposition is in the air. It might be expressed as just 'loving/hating', or as 'we together', or as 'deeply united'. The zero-term is appropriate where the sense of unity has gone beyond the deepest emotional feeling into the deep peace of Matte Blanco's fifth level. I will return to this later in the context of mysticism.

Suppose now that the son is often tormented by the daughter. The son's assessment might be 'She is always attacking me'. In his unconscious symmetry yields 'and I am always attacking her'. The corresponding propositions may describe the daughter's assessment too. The overt aggression of the daughter may indeed be linked to a covert emotional aggression by the son, a provocation not immediately obvious. The theory of bi-logic encourages the observer to speculate that these unconscious deductions are in place, and thence to see realities that might otherwise not be considered. In this situation, compounding the first propositions and their derivatives gives 'we are always quarrelling', and that may certainly be a more complete account of the emotional situation than regarding it as one of unilateral aggression. 'We are always quarrelling' leads, as a one-term proposition, to its opposite, 'We are always happy together' and this may express another important truth.

Within the same overt situation there may be another emotional conflict at work. Suppose that the father often intervenes to rescue the son from being tormented by the daughter. In each intra-subjectively this may be represented by a three-term proposition, 'F rescues S from D'. Unconsciously this may be rotated to produce, 'S rescues D from F' and 'D rescues F from S'. Karpman[8] in a well-known paper explored this triangular situation in the context of social work — the social worker rescues the victim from the persecutor. He argued that the intervention may be experienced in quite a different way, in which indeed the rescuer is the persecutor and at a deeper level the rescuer is the victim, with the other two actors taking up the positions predicted by the law of rotation. Compounding the three (three-term) propositions, gives us a

[8]   Karpman (1968).

one-term proposition, 'We three are rescue-triangling'. I use the term 'rescue-triangling' for lack of an appropriate word to describe the 'game' in hand. Note that in this particular game the mother has no role. If we were to assign her one, we might be involved in the complexity of a four-term proposition, such as 'the father rescues the son from the daughter to annoy the mother'. It may be that four-term propositions properly represent some situations, but my own guess is that they are usually composite. The above situation may be better represented by two propositions: 'we three are rescue-triangling" and "we rescue-triangle to annoy mother'.

The somewhat jejune examples I have discussed have been chosen for the sake of clarity to illustrate the theory of symmetric logic, as the logic of the unconscious. Of greater psycho-analytic interest would be a discussion of the rotation of the Oedipus and Electra triangles, and the significance of the deduced propositions: but that would go into uncharted territory beyond the scope of this chapter. Since emotion comes from the unconscious, symmetric logic will be at work in all emotional situations. The asymmetric logic of the conscious mind will generally be discernible also and the resulting bi-logical structures are the rich stuff of human life wherever it transcends expression in purely factual and empirical discourse.

The Interplay of *Same* and *Different*

Matte Blanco gave great prominence to the set-theoretical aspect of symmetric logic in unconscious process. I have already illustrated this in the discussion above of authority figures. Matte Blanco formulated this in his law of generalisation.[9]

> The system Unconscious treats an individual thing (person, object, concept) as if it were a member or element of a set or class which contains other members; it treats this class as a subclass of a more general class, and this more general class as a subclass or subset of a still more general class, and so on.

With the help of this 'law' the working of the two logics can be seen in a vast range of human activities and not only in the dynamics of human inter-relationships. The law of generalisation suggests that the registration of 'sameness' may be linked to symmetric logic, while the registration of 'difference' is linked to asymmetric logic. Consider, for example, somebody collecting and classifying butterflies. In one swing of the net two butterflies are caught. They turn out to be two distinct individuals of the same type and both classified as, say, Tortoiseshells. For the purpose of classification their slight individual differences are ignored and

[9]   Matte Blanco (1975), p. 38.

they are assigned to the same class or set and they are seen as the same. Seeing two different things as one and the same is an exercise of symmetry. The collector is thinking 'Butterfly A and butterfly B are the same' although they are different individuals. In the next swing of the net two butterflies are again caught, but in the respects which interest the collector they are different. One is a Tortoiseshell, the other a Red Admiral. The collector thinks 'Butterfly C is not the same as butterfly D'. This asymmetric distinction is not made symmetrical.

Classification therefore is a bi-logical activity. It involves making certain distinctions and ignoring certain other distinctions, asymmetry and symmetry. Butterflies may be classified according to the established scientific taxonomy, but of course they could be caught and classified in other ways. A bizarre collector can be conceived who classified them only by their predominant colour, or weight, or wing-span. To one such, two butterflies of the same weight would be symmetrically treated as the same, even if they were of different species, while two of the same species but of different weight would be counted as different. This sounds rather absurd, but illustrates an important point. Symmetrising (i.e. treating two different things as the same) always proceeds on the basis of a choice of what is to be ignored and what is to be kept in the picture. The curators of a zoological collection may be assumed to assess their animals, like the butterfly collector, in the customary taxonomic categories. The person charged with supplying food for them might adopt a quite different system whereby to him a grizzly bear is equivalent to a lion if they eat the same amount of meat daily, while the wild cat (which the taxonomist might see as belonging to the same class as the lion) may be equivalent to a jackal.

The different activities and interests of humanity can be analysed from a bi-logical perspective in terms of what is registered as 'the same' using symmetric logic, and what distinctions are preserved. Students of bi-logic have applied this in a number of fields. For example, an accountant charged with assessing the estate of a deceased rich person evaluates his possessions financially. An art collection may be deemed, for this purpose, equivalent to a country estate or an impressionist work of art to a dozen Rolls Royces. The distinctive features of each item are left out and only the financial value remains creating equivalences between very different items. Legal systems, at least in the post-feudal West, incorporate the principle of equality under the law between people whatever their age, sex, status etc and thus emphases that two different people are of equal legal value. Many such examples could be given and the hidden symmetric logic disclosed. In all of them it is important to note that though symmetric logic is at work this is not necessarily a consequence of unconscious process, since some symmetries are deliberately created in consciousness. However, since symmetric logic is the kind of reasoning natural to the unconscious, it is likely to apprehend such symmetries and perhaps invest them with meaning. Perhaps Mammon becomes a god because its underlying symmetry (everything may be equated with a certain financial value) offers a simple way of coping with reality.

Asymmetric logic is believed to have first been analysed by Aristotle and indeed is often called Aristotelian logic. It is not surprising therefore that it links with other features of his philosophy, a philosophy which is in sharp contrast with the philosophy of his master and teacher Plato. I shall move on now to suggest that symmetric logic has affinities with Platonism and indeed provides a key to understanding much that is deeply uncongenial to the spirit of scientific modernism. Aristotelian logic, being asymmetric, is suited to the drawing of distinctions, the analysing of a thing into its parts, and to the accurate handling of factual matters. Aristotle's philosophy involved such distinctions and led to the distinguishing and listing of the contents of the universe. It proceeded by definitions of genus and species, narrowing these to discriminate differences. It thus prepared the ground for the Renaissance and Enlightenment development of modern science and technology. As a broad generalisation Aristotle moved from the general to the particular in the search for understanding. In contrast Platonic philosophy moves from the particular to the general. Any thing is what it is because it is an expression of an ideal Form and forms exist, not in the world of matter but in a transcendental world of thought. Whereas Aristotelian philosophy tends towards the typically modern view that what is real is what is material, Platonic philosophy sees reality as imparted to matter by the Form that shapes it and the Form is what is genuinely real. Furthermore the Forms themselves are increasingly real as they are increasingly general. Ultimate reality is found in the Form of the good, the One from which all other (less real) realities emanate. Aristotle looked for distinctions and differences — the essence of asymmetric logic. Plato looked for likenesses and samenesses, the essence of symmetric logic. Aristotle's thinking particularly laid the ground for modern science. Plato's successors developed his philosophy in a religious and more specifically mystical direction.

In expounding Plato's Forms for a popular audience, it is customary to take up his illustration from the tenth chapter of the Republic.[10] There he suggests that any particular bed is a bed because it participates in the form of the ideal Bed. Particulars participate in the Forms and the Forms are instantiated in particulars. It is, however, by no means clear that this

---

[10]   Plato (1955), pp. 371-3.

illustration was meant to be more than an illustration. What concerned Plato was not so much how a bed is a bed, but how humanity is drawn to beauty, truth and goodness. The human body can be beautiful, but what one may be drawn to is ultimately beauty itself. I propose to connect this with psycho-analytic thinking. In my discussion of fathers and authority figures I argued that every possible father can be present to the unconscious in a relationship that apparently is merely with a line-manager. It is as though the Form of Fatherhood were present, a Form first known in the particular case discussed, as oppressive. Likewise a motherly figure may be felt to be a renewed experience of one's actual mother and in her all motherliness may seem to be present. She instantiates the Form of the Mother. It is note-worthy that Kleinian psycho-analysts use for this the term 'the Breast', usually spelt with a capital letter and surely echoing the Platonic notion of an ideal Form.

In connecting Platonism and psycho-analytic theory, I am doubtless somewhat revising the former, certainly the customary perception of it. I am proposing that in matters of deep importance to the human soul an unconscious logic is at work that coheres very much more with a Platonic understanding of things than it does with the empiricism of the science of the last few centuries. The things that most deeply move us are few in number and perhaps all represent a certain primal good that might be equated with the ultimate Platonic Form.

Symmetric Logic and Mysticism

I shall now take up the suggestions already made that the unconscious and the mystical are closely linked in a way particularly revealed by the concept of symmetric logic. The word 'mystical' is used in many senses, and there is no satisfactory definition of it, although a search for direct experience of God is generally comprised in it. I shall restrict the term to the theological stream generally described as apophatic mysticism. A leading feature of this is, paradoxically, that direct experience of God in his essence is impossible and visions and special revelations are regarded with suspicion. The movement towards God is attempted by stripping away sensation, imagination, discursive reason and, while in an entirely conscious and attentive state, attending to nothing, the void from which or in which (even as which) God may make himself known.

While it would be inaccurate to describe all such mystical theologians as Platonists, Platonism has had an immense influence in this tradition. It may be seen as originating with the first century Jewish thinking, Philo, as stimulated by the pagan philosopher Plotinus and with the early Platonistic Christian theologians, Clement of Alexandria and Origen. It continues in the East with Gregory of Nyssa and in the West with Augustine and finds particularly clear expression in Dionysius the Pseudo-Areopagite. In Eastern Orthodox theology it is the dominant trend, while in the West Dionysius had an immense influence through the Middle Ages, upon the Victorines, Meister Eckhart, Bonaventure, and the anonymous author of the *Cloud of Unknowing*, perhaps culminating in the work of Nicholas of Cusa (which incidentally influenced Matte Blanco's thinking very strongly). In this tradition the approach to God is made by attempting to clear the mind of all particulars and all imagination and all symbols of God. It is an attempt to turn the mind from anything confined by space and time, to put all such under 'a cloud of forgetting' and turn instead towards 'the cloud of unknowing' beyond which, or in which is God. And God is not here to be though of under any possible shape or conception, but as mere existence or being. This movement of the mind is a work of symmetric logic wherein all the distinctions between one thing and another that characterise normal conscious thinking are abandoned, even such distinctions as might be thought to characterise God over against his creatures. It seems to me clear that these mystics are attempting consciously (in their consciousness) to enter the deepest unconscious. This view was certainly that of Bion (a very influential psycho-analytic writer) as well as of Matte Blanco, and also of Jung (the great heretic in the psycho-analytic world). One important modern writer, on mystical literature, Johnson, also supports it and certainly most of the mystical theologians, Plotinus and Augustine for example, write of going into oneself to find God, while Eckhart leads his readers towards what he calls 'the innermost'. While it seems a difficult proposal to prove, it seems very reasonable to equate the state of being commended by the mystical theologians with the conscious attainment of the stillness of the deepest unconscious as Matte Blanco describes it.

If it be allowed then that the God of the mystical theologians and philosophers is either the same as the deepest Unconscious, or at least is encountered in the deepest Unconscious, the theory of symmetric logic provides further interesting lines of inquiry. One of these concerns the ancient discussion of 'the One and the Many', for in the light of bilogic, symmetry perceives reality as one, whereas asymmetry perceives it as many. On the one hand stands the philosophic tradition of monistic idealism (of which perhaps Spinoza is the most extreme exemplar), on the other of empiricism, with its understanding of the world as an aggregate of discrete entities and facts. In theological terms there has always been a tension between the concept that God is essentially all-in-all, and on the other that in creation a new reality (or realities) stand over against 'Him'. In the early centuries of the Christian church a particular problem arose over the unity of the divine itself. On the one hand the

Jewish Scriptures (the Old Testament) emphasises that God is one. On the other hand, the New Testament revelation hinted at least that in Jesus a second divine reality was at work, and even a third in the shape of the Holy Spirit. The influence of Greek philosophy reinforced the notion that God was one, while perhaps the polytheistic background of religion and of the mystery cults made easier a recognition of the Three. Centuries of fierce disputation were resolved in the formula of the doctrine of the Trinity, three beings in one being, a paradoxical assertion of an intentionally inconceivable concept. The Athanasian Creed asserts for example that 'Such as the Father is, such is the Son, and such is the Holy Spirit', and explains that each is uncreate, incomprehensible, and eternal, yet there are not three uncreates, incomprehensibles or eternals, but only one. Yet again, the three are not to be confused with one another, for they are both absolutely distinct from one another, yet absolutely one. Naturally the sceptical have always regarded this doctrine as pure nonsense, since it is clearly counter to reason, to asymmetric logic that is to say. The doctrine is, however, equally clearly acceptable to symmetric logic, a confirmation, I would argue, of the essential identity of the deepest unconscious and one conception of God.

The argument of this chapter is that the human mind uses two logics. The scientific thinking of the last few centuries has greatly elevated the status of the rational and empirical and this has been essential to its methods and to its success. The negative side of this development has been the relative neglect of much that is essentially human. A purely objective approach to reality ignores human subjectivity and leads to a soulless and incomplete understanding. The arts and religion may become regarded as mere diversions without a claim to offering insight into truth. In such matters symmetric logic is prominent and the theory of bilogic enables one to trace its footprints. In so doing a greater respect may be accorded to things essential to true human prosperity and a rift that can be traced back to the distinctive emphases of Plato and Aristotle may be better understood and, perhaps, overcome.

Chris Clarke

Both/And Thinking

Physics and Reality

The Limits of Current Science

Science has steadily risen in prestige over the last four centuries, to become the dominant way of knowing in the world today. It represents the pinnacle of reliability and certainty. Once something is 'scientifically proved' then all further doubts and questioning are laid aside — though, paradoxically, for the great majority of people the actual operations of science remain a mystery. Science seems to satisfy a craving for security and certainty that many feel today, in a world where they are increasingly powerless, facing an increasingly unknown future. Science, it is thought, gives a picture of a universe where there is no ambiguity; where every object has its precise place and its precise properties; where there are no limits to the power of the human intellect. It seems to give a reassurance that something, somewhere, can be relied on. In comparison, other ways of knowing are denigrated as 'subjective', 'poetic' or 'fanciful'.

This appeal to science for a sense of certainty extends, moreover, to the more powerful parts of our human experience: to consciousness, to religious/spiritual experience and to the sort of numinously charged exceptional experiences that lead to a fascination with parapsychology. Here the desire for certainty is either expressed through forms of religious dogmatism, fuelling the rise that we now see of the conservative elements within all major religions; or it again appeals to science for explanations. The consequences of the latter are apparent in the popular writings that abound in our bookshops: either the authors resort to invented pseudo-science to explain these phenomena in terms of universal 'energy fields' that can accommodate everything but in the end explain nothing, or else they dismiss the phenomena by reducing them to neuropathology and selfish genes. All this is misguided. It stems from a fear of the unknown, and fear is always a bad counsellor. The desire for reassurance has degraded religion, grossly distorted the popular attitude to science, and has started to distort the funding and practice of science itself.

The bad news is that this makes us unable to grasp the profound dangers now facing society and the planet, whose origins are basically spiritual, in the sense of concerning the totality of our values and ways of perceiving. By looking for instant scientific and religious 'fixes' for what are at their root spiritual problems, we are locking ourselves into an increasingly dysfunctional relationship with each other and with the earth.

The good news, which I want to explain in this chapter, is that we now can start to reach a different form of understanding — but only by firmly rejecting this popular appeal for certainty from a simplistic conception of science. Science (and I am here talking about the particular social activity that goes by this name, not about abstract principles) is both much more subtle and much more limited than one might suppose from watching TV documentaries. It is one way of knowing among many, and if science is to enhance life in all its aspects, and for all social sections and for all species, then it is imperative that we widen our vision to see its connectedness with these other ways. We might then still want to call the enlarged vision 'science', but we must be clear that I am advocating a true revolution in relation to science as it is now practiced and understood. I will be explaining how science, and more generally the rational analytical approach of which science is the best example, points towards other ways of knowing and enables us to understand how these all fit together. We do not have to dismiss other ways as merely poetical; nor, on the other hand, do we have to abandon science entirely in favour of a form of mysticism about which we must, as Wittgenstein at first suggested in his early *Tractatus*, 'be silent'. Instead, we can become able to see the whole landscape of human enquiry through both rational eyes and intuitive/relational eyes.

Extending the Foundations of Science

Certainty, whether derived from religious dogmatism or misunderstood science, rests on a particular constellation of ideas about reality, logic and truth, a constellation that has for centuries formed the foundation of science. We must start by questioning this, if we are to extend the realm of current science. It is assumed that there is a single real world; that it is adequately describable by propositions obeying traditional rules of logic, and that our theories can become closer and closer to the truth to the extent that they match the structures of the real world. (In philosophy these ideas are collectively referred to as 'the correspondence theory of truth'.) This is an immensely alluring conception. Note how its three terms seamlessly reinforce each other. Why is truth to do with matching the world? Well, that's only logical. What's great about logic? It's the only way to be sure of grasping the truth. What is truth, anyway? It's about matching the real world. This conception stands or falls as a whole. We are so used to this way of thinking that we assume that it is simply how the world is. But this conception is just one particular cultural form of one particular part (the so-called propositional part) of the cognitive apparatus of the human species: I suggest that when we think we are talking about how the world is, we are just talking about *homo sapiens propositionalis Graecae*.

In order to unpack this claim, I will first discuss the way in which only a part of our cognition, our ways of 'seeing', is recognised by this notion of truth. Isabel Clarke in her chapter has described how, on one influential conceptualisation of the mind, our way of 'seeing' is composed of the interplay between two main subsystems of our mental apparatus: the *propositional,* to do with language and focussed reasoning, and the *relational,* to do with our relationships with the rest of the world, immediately communicated through all our senses. Under stress this interplay can break down, reminding us that the human condition is a 'balancing act' between two extremes: on the relational side, complete identification with, and loss of boundaries in, the immense presence of the world around us; and, on the propositional side, complete domination by the rational analysis of world.[1] Human growth, even survival, requires our ability (both as individuals and as societies) to navigate effectively in the straits between these polarities. Both must be present. On the propositional side, the power of rationality is indisputable. On the relational side, as David Abram argues in his chapter, the very substance of our lives depends on our intimate sensual connection with the beings of the world around us; and modern mystics would agree as to the importance of opening our boundaries relationally to the whole from which we draw our being (though they might name and construe it differently from the way I am doing here).

The scientific way of knowing, with its correspondence theory of truth, is dominated by the 'propositional' subsystem of our mind. This way of knowing can say nothing about *what it is like* to see the sun rising in the morning, or to smell a rose garden on a summer's evening. The

---

[1]   This is related to what Rodney Bomford, also writing in this volume, has — from a different perspective — called the Scylla and Charybdis of human experience, Bomford (1999).

best it can do is to describe, in functional terms, how the sun and the roses influence our behaviour and speech, but it thereby evades and denies the essence of our inner experience, which lies in the relational subsystem. Science is incapable of penetrating the experiential aspects of our relations with the world. Consequently, our exclusive obsession with science, and the Western philosophy that underpins it, widens the gulf between the relational subsystem, with its ability to allow us to relate to and identify with the world, and the propositional subsystem, whereby we can understand it rationally. The polarity is turned into a dichotomy. Either we are dominated by the rational, using logical reasoning and regarding reality as no more than the objective world revealed by classical science; or we are irrational, loosing hold on reality, falling down a slippery slope into madness. From this follows the pathologisation of the mystical and the consequences for society described by Jennifer Elam earlier.

For clarification, I would stress here that the roles of the relational and propositional subsystems are not symmetrical. For while it is possible, in extreme mystical states for example, to be located entirely in the relational, it is not possible to be located entirely in the propositional — because it is the relational that gives substance to our awareness, and without the relational there can be no cognition at all. This reinforces the point made by David Abram in his chapter, that it is the sensual world that is foundation for all our other 'worlds'.

As a practical strategy, we urgently need to find in ourselves our connectivity with the world (a connectivity that we are born with) and allow this more and more to fill our lives. In addition, however, we need to build on our existing intellectual achievements by widening our thinking so as to heal the gulf that has opened. This is the aim of the present chapter. I will describe a new understanding of the nature of thinking — an extension to what is sometimes called 'both/and thinking' — which is not restricted to rigid boundaries, and which enables us to be rational and mystical at the same time. A key role in this understanding is played by an enlargement of logic to something much wider than the correspondence theory of truth discussed above.

To grasp the widened vision, we will need to accept that *there is no single all-embracing absolute propositional truth.* Instead, propositional truth is *contextual,* depending on the particular occasion, not absolute. The idea of absolute propositional truth is contrary to our experience and, as I will shortly illustrate, it is contrary to the implications of modern science. It is a hang-over from the early days of philosophy whose time has passed. Widening our horizons enriches us not by converging to a final truth, but by enlarging the range of truths.

It is crucial here to understand the difference between contextual truth and relativism. The latter claims that there is no such thing as truth at all and that all ways of construing a single world are equally valid, a rash claim that is just as much against the evidence as is the traditional claim that there is absolute propositional truth. My argument will be not for relativism, but for the *pluralism* of a contextual truth. In pluralism there are a multiplicity of intermerging worlds whose limited, local, truths fail to mesh into a single propositional unified global truth. This is not to say that there is no truth at all, however, because of the role of the relational and propositional subsystems. On the relational side, we can find, in powerful sensory terms rather than propositional terms, a guidance that can proclaim the validity or invalidity of our path. There is an intuitive sensory grasp of a greater dynamic 'truthing', a verb rather than a noun, which escapes from the confines of the propositional and knows neither logics nor bounded worlds. Reaching an adequate balance for the current situation of humanity requires the interplay between contextual logic and non-logical truthing, corresponding to a seamless interplay between the propositional and relational subsystems of our mind.

Logic and Language

You have before you a book full of words. Does this mean that we are already imprisoned in some sort of logical structure? Logic is certainly bound up with language, as I have already implied by associating both logic and language with the propositional system. But language can do two things, intermingled. It can *encompass* the propositional, and it can *point to* the relational. Corresponding to these two, in Western writing there tends to be a distinction between the language of philosophy (including Natural Philosophy, i.e. science) and the language of the mystics. The former draws on the propositional subsystem, tending towards precise definitions and closely delineated reasoning, and strives to be 'logical'; the latter, drawing on the relational system, often employs negative assertions and paradox, amongst a wide range of literary styles, and sometimes either implicitly or explicitly repudiates logic.

We might take as examples Rumi's words:

> I am neither Christian, nor Jew, nor Parsee, nor Moslem. I am not of the East, nor of the West, nor of the land, nor of the sea ... my place is the Placeless, my trace is the Traceless ... [2]

and Nicholas of Cusa:

---

[2]   Rumi (1898), quoted by Soelle (2001), pp. 65–6.

> [*Addressed to God*] ... I have discovered that the place where you are
> found unveiled is girded about with the coincidence of contradicto-
> ries. This is the wall of paradise, and it is there in paradise that you
> reside. The wall's gate is guarded by the highest spirit of reason, and
> unless it is overpowered, the way in will not lie open. Thus, it is on
> the other side of the coincidence of contradictories that you will be
> able to be seen and nowhere on this side.[3]

And yet, as Rodney Bomford has demonstrated in his chapter, there is a
form of logic, the so-called *bilogic* developed by Ignacio Matte Blanco,
that can shed much light on the nature of mystical experience. This is a
vital example of the sort of enlargement of logic that we are looking for.
Before examining this, and other examples, however. I need to make
clearer just what I mean when I talk about 'logic'. To explain this, I will
give a brief historical sketch of how the more rigid sort of logic was
established, and how it has started to be generalised.

   Classical, restricted logic is a notion that goes back to Aristotle's *Prior
Analytics*. Here he discusses what patterns of reasoning are *valid,* that is,
will lead to correct conclusions, if the premises are correct, as opposed
to patterns of reasoning that are *invalid* and may lead to incorrect conclusions from correct premises. He assumes that, if we think about the
matter, it becomes obvious which patterns of reasoning are valid and
which invalid. And his intention is to lay down the principles that we
ought to use for valid forms of reasoning. His logic is prescriptive, not
descriptive. This is the sense of logic that dominated classical times.

   Underlying Aristotle's approach is an implicit theory of truth and
knowing, which he shares with his tutor Plato, despite their famous differences, and which is essentially a version of the correspondence theory of truth, described earlier. It is assumed that there is a single truth
about the world; that this truth is, at least in part, unveiled to the human
intellect when we apply it to the world; and that the truth thus revealed
conforms to the logical structures of the (Greek-based) language in
which we express it. The Greek for 'truth', *aletheia*, expresses this
through its etymology of 'un-concealed'. Similarly, the Greek *theorem*
derives from *theoria,* meaning vision, indicating the way in which a
mathematical proof was supposed to operate by revealing directly to
intellectual vision the way that things are.

   This theory of knowing is directly challenged by mysticism. First,
while the mystics often imply that they have immediate contact with
Truth[4], they are often, as we have seen, insistent that it is *not* conform-
able to the logic of normal language, and are driven to use words in a

---

[3]   Nicholas of Cusa (1997), pp. 251–2, an extract from his *The Vision of God*.
[4]   Though Dionysius states that the divine is 'neither knowledge nor truth',
       Dionysius (1987). p. 141, quoted Soelle (2001), pp. 66–7.

way that disrupts their normal relations and meanings in order to try to point beyond them to what they cannot express. Second, although each mystical tradition might be clear as to the unity of the truth that is being revealed specifically through it, the more recent scholarship of mystical traditions — as Ferrer shows in his chapter — indicates that this truth is *not* unitary over traditions, but differs in essential ways from one tradition to another. We see here in historical, social terms the gulf between different ways of knowing that I have already described. Aristotelian logic must be widened in order to bridge this gulf.

In modern times the first steps towards doing this were taken when interest shifted from logic as a *prescription* for how people *ought* to reason, to being a *description* of how people *actually do* reason. The shift took place around the time of the anthropologist Levi-Bruhl,[5] who noted that many very intelligent indigenous people reasoned in ways that were clear and systematic, but not according to Aristotle, and as a result they reached conclusions that a Westerner would think strange. He was tempted to call their reasoning an 'alternative logic', but was unhappy with this term because it clashed with his own idea that there was only one right way to reason. His successors had fewer qualms, however, and were happy to introduce the idea of alternative logics.

Before long experimental psychologists[6] started examining how people actually reasoned, and discovered that almost no one stuck to Aristotle's prescription for much of the time. What is happening here? Is it that we are just not very competent, trying to reason properly but simply making mistakes, like wrong stitches in knitting? Or are we in fact following regular principles embodied in our psychology, perhaps evolved for getting to conclusions quickly that work most of the time (e.g. following 'rules of thumb'), but using a form of logic different from Aristotle's?

In parallel with this, mathematicians began to formalise the processes of logic, out of which began to emerge a new genre of *mathematical logic* that was no longer expressed in terms of ordinary language. This started with workers such as A.N. Whitehead, trying to uncover a single fundamental logical system that underlay all reasoning, and mathematical reasoning in particular. What actually emerged, however, was a wide variety of formal schemes of deduction, each with its own non-standard concept of 'truth'.

In the course of this, mathematical logic began to be distinguished from traditional logic based on ordinary language. The divergence between the two can be illustrated in the case of the idea of a *proposition*.

[5]   Levy-Bruhl (1985).
[6]   Wason and Johnson-Laird (1972).

In natural language a proposition is a statement that might (at least in the simplest cases) be true or false; such as 'this book weighs 500g' or 'all sycamore trees bear seeds with an attached wing'. This logic is concerned with combining propositions with conjunctions such as 'and' or 'or' so as to form *compound propositions*. And in such cases the word 'logic' is often used in a restricted sense to mean the *set of rules* for determining the truth or falsity of compound propositions in terms of that of their components, and determining the equivalence or otherwise of various compound propositions. The same is true of mathematical logic, but everything now becomes abstract. A 'proposition' is just an abstract entity in a mathematical system, just as numbers are, 'conjunctions' are just symbols analogous to the '+' and '−' of arithmetic, and 'true' is just an abstract quality possessed by propositions, analogous to whether a number is positive or negative, having no meaning in itself. It such a situation, alternative logics can be spun at will in the quest for interesting abstract structures, unfettered by considerations as to what the world is actually like!

Mathematicians are often surprised when their constructions turn out to be useful in the world, as happened when quantum theory showed that logic had to be altered to a different system even in the supposedly down-to-earth area of physics. I want to describe this in the next section, partly because it has some analogies with the logic of mysticism or of the everyday human mind, but also because of its cultural importance in bringing home to the intellectual world the demise of Aristotelian logic. Once we realise that the Aristotelian theory of knowledge is not inevitable, then we start to see a whole spectrum of different ways of knowing, with different logics, which is relevant not just to peak mystical experiences, but to physics, psychology, education … indeed to the entire relationship between humanity and the world.

Before continuing, let me once again stress that modern logic is *not* 'how we ought to think'; indeed it is not a part of thinking at all. Logic and mind are different: one is an abstraction, the other is Being. But they still can be descriptively related. We need to think of logic as being like the mathematical models of air circulation that underlie weather forecasting. The neat weather maps with their smooth isobars and fronts decorated with formal symbols are a complete idealisation which we do not expect to correspond precisely to whether or not it will be raining; but we are happy if, on the whole, there is a helpful correlation between the maps and what actually happens. It is the same with logic as an abstract description of the sort of things that can happen in the human mind — and hence a description of the sort of concepts that we can form about our world.

In the next two sections I want to give as examples two sorts of non-Aristotelian logics, namely quantum logic and the logic of Matte Blanco described by Bomford in his chapter. I am not claiming that either of these is 'the' replacement for Aristotelian logic; indeed, it would be in keeping of the spirit of this enterprise to suggest that there is no such single replacement. But they will serve to indicate the sort of loosening up that I have in mind.

Quantum Logic

Quantum logic is simply a logic in which the rules for the truth or falsity of compound propositions are given a particular modification from Aristotelian logic, which I will describe with an example.

Suppose I go into a café and note from the menu that I can have sausages, and either fried eggs or poached eggs. I accordingly order sausages with fried eggs, to be told that this is unfortunately unavailable; I therefore change to sausages with poached eggs, only to be told with great regret that this is also unavailable. Yet when I direct the waiter to what is written in the menu, he affirms that 'sausages, and either fried eggs or poached eggs' is nonetheless available. I am confounded, because my logic contains the rule

if A, B and C are propositions, then the compound proposition 'A and (B or C)' is equivalent to '(A and B) or (A and C)'.

(This is called the *distributive law* of formal logic.) I was taking A, B and C to stand for 'sausages are available', 'fried eggs are available' and 'poached eggs are available', respectively. In that case (allowing a little linguistic flexibility) 'sausages, and either fried eggs or poached eggs are available' should be equivalent to 'sausages and fried eggs are available, or sausages and poached eggs are available'. The waiter, on the other hand, seemed to be using some different logic, in which this law did not hold.

We might further suppose that my suspicion was confirmed when I ordered 'sausages with either fried eggs or poached eggs' and was served with sausages and scrambled eggs, being assured by the waiter that 'fried or poached' obviously included all other sorts of eggs, some of which were available.

The waiter was in fact using quantum logic, a system in which all the usual laws hold except for the distributive law. Its characteristic is that it is usually not possible to assign values of 'true' or 'false' in a consistent way to all propositions. Instead one moves from one limited set of propositions to another, keeping the assignments consistent within each set, but with there being no universal concept of true or false. The most famous example of this in quantum mechanics is that of 'wave particle duality'. This refers to the fact that, speaking rather imprecisely, we can either describe the sub-atomic world in terms of propositions about particles, or in terms of propositions about waves. In each area there are consistent assignments of 'true' and 'false', but one cannot extend these over a single domain covering both waves and particles.

Logic thus becomes context-dependent: what is the case depends on whether the context is one of waves or of particles, of eggs or of sausages. Context-dependence occurs when the concepts involved have a certain fluidity, which one does not expect of restaurant menus, but which is realised once one starts to stray from the area staked out by the language of practical crafts, whether we stray towards the interior experience of mysticism, or towards the extreme world opened up by the instruments of atomic physics.

Bilogic: Context and Structure

The bilogic of Ignacio Matte Blanco described by Rodney Bomford in his chapter represents a more radical generalisation of logic than quantum logic. I shall give a short summary here drawing out the features I need for my own discussion: the reader is referred to Bomford's article for more detail.

Like quantum mechanics, bilogic is context dependent. This is seen in Matte Blanco's law of generalisation:[7]

> The Unconscious treats an individual thing (person, object, concept) as if it were a member or element of a set or class which contains other members; it treats this set or class as a subclass of a more general class, and this more general class as a subclass or subset of a still more general class, and so on.

In other words, the logic depends on a nested sequence of classes which form *contexts* for the operation of the Unconscious on different occasions.

Within each context, bilogic uses rules of inference of two forms: either ordinary Aristotelian rules, or rules from a special new sort of logic called symmetric logic. The latter works with a fundamentally different structure from classical logic by including the rule that every proposition can be turned round into its converse. To consider a hypothetical example, in symmetric logic one can deduces that

'Dragons are dangerous things'

is equivalent to

'Dangerous things are dragons'

---

[7]   Repeated here from p. 137 above.

From this one might argue that, because my acquaintance Peter seems rather threatening, and hence dangerous, Peter is a dragon, and so presumably he can fly, breath fire etc. The connecting word 'is/are' here is not the usual 'are' of 'all sycamore seeds are formed with an attached wing'. Instead, it is rather like the relationship of *association* which is basic to the technique of psychoanalysis. A conventional analyst would investigate a patient's claim that Peter can fly by uncovering an association between 'Peter' and 'dragon', without postulating a whole underlying logic. The advantage of moving from the straightforward analytical tool of association to the more elaborate idea of a symmetric logic as a model is that it unifies a whole range of properties of the unconscious. Replacing conventional logic by a context dependent logic having a different formal structure is a way of representing, as an idealised model, a process that has its own regularities but is much more fluid than the processes of classical logic.

Although his writing is dependent on a Freudian approach, the idea is much wider than this, and it seems to fit as well, or even better, with the subsystem model. Symmetric logic points to the relational subsystem, in the way that language can point to this area without precisely describing it, as I described above. We see this because, when the relational system becomes decoupled from the propositional, then the person behaves *as if* they were governed by the interplay of the two components of bilogic. *Bilogic is thus a model that connects the two sides of the 'gulf' in the human condition.* The feeling of "infinity" associated with the relational subsystem fits perfectly with the way in which infinity emerges as an inevitable consequence of bilogic, as has been described by Bomford.

The Science of an Enlarged Logic

I now want to return to enquire how the Aristotelian theory of knowing is altered by the perspectives of quantum logic and bilogic, with particular emphasis on the former. In order to do this, I first need to give a feel for what quantum theory is like, in comparison with other scientific theories.

If there is such a thing as a 'typical' scientific revolution (an idea that has been increasingly contested) it might follow the pattern that unfolded with the theory of gravity in the early years of the twentieth century. Initially there was unease about the traditional theory, that had been in place since Newton in the seventeenth century. On the experimental side, the details of the orbit of the planet Mercury were minutely different from what was predicted by Newton's theory; and on the theoretical side it was unclear how to combine Newton's theory with Einstein's newly developed theory of Special Relativity, dealing with the mechanics of bodies that moved very fast. Then Einstein, motivated mainly by the second of these problems, produced a radically new theory of gravity (namely, General Relativity). This not only fixed the existing problems, but it neatly incorporated Newton's theory as an approximation to General Relativity, and it opened up a whole new field of investigation that otherwise would not have been thought of — in due course including black holes and the big bang. Thus the sequence of events was

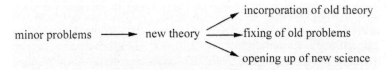

The case of quantum theory was significantly different from this. To begin with, the problems with the old theory were not minor, they were catastrophic. For example, one problem was to calculate the colour of a body glowing with heat — why does a skewer heated in a gas flame glow *red* hot, while the steel from a blast furnace is *white* hot? Attempts to do this not only failed to calculate the colour, but they predicted that a hot body should emit infinitely much radiation! Secondly, while the new theory, quantum theory, successfully solved these, and many other, major problems, it did not fully incorporate the old theory. Instead, it sat alongside it, and required physicists to switch adroitly from one to the other in order to get sensible results. It certainly opened up new science (atomic power/bombs, computers, cosmology, the foundations of chemistry, astrophysics ... these all now depend essentially on quantum theory) but it kept achieving its successes at the same time as introducing bizarre assumptions and violating all the rules of proper mathematics. It was like a mad genius whose eccentricities were, however, forgiven for the sake of what it created. So, while most physicists got on with exploring the successes, others continued to try to tame this genius and reformulate the theory in such as way as to make science a consistent whole.

This brings us to the recent work that makes contact with context-dependent logic. Over the last twenty years one of these reformulations, called the histories approach, seemed to be achieving the sought for consistency. The idea is in essence quite simple. In classical physics the universe has a definite state at every time and rigid laws determine how this state changes from moment to moment. In the histories approach one looks for a valid sequence of propositions (a *history*) that describes the process of the universe, or an aspect of it, in more or less detail. For example, 'First there is a cloud of gas; then it forms galaxies; then they produce stars; then at least one of them produces planets ...'. Or 'In a certain laboratory there are two protons travelling towards each other with a speed of ... Ten microseconds later there are ...' There is a shift from a (classical) system of laws determining progressive change in a complete specification of reality to a system of constraints on broad scenarios of observations. The mathematical formalism of quantum theory is used to produce a *probability* for each history that one considers. Later Hartle[8] generalized the idea of a history from a simple sequence of propositions referring to different times, to a 'network' of statements referring to different times and places. E.g. 'meanwhile, in another part of the galaxy ...' Finally Chris Isham and Jeremy Butterfield[9] interpreted the different times-and-places as a succession of contexts with a quantum logic holding at each one, and reformulated the idea as what I am here calling a *context dependent logic*.

A lot of the conventional scientific picture is retained; in particular in the process of interpreting a natural language proposition, such as one of the examples just given, into a proposition in the sense of mathematical logic: an object within the mathematics of quantum theory. Indeed strictly speaking a 'history' in the sense of this approach refers not to a sequence of natural language propositions, but a sequence of mathematical objects referred to as propositions because they obey the laws of mathematical logic. The interpretation of a natural language proposition as a mathematical proposition involves a great many assumptions about what the universe consists of and how these are mathematically represented. But there is nonetheless a decisive shift from the picture of a machine to the picture of a discourse involving a fluid logic.

The generalised histories interpretation has great elegance, in that it pares down the quantum theoretical account to its bare bones without losing any of its predictive content. Much of the baggage that was brought into the early interpretation, including the mysterious 'collapse of the quantum state', is swept away with the rejection of the idea that the universe has an absolute state at each time.

Unfortunately, however, the sought-for consistent picture is still not achieved. The aim of a physical theory is to match, more or less, the whole range of phenomena within an area where it is supposed to apply. Each phenomenon observed should correspond to a description within the theory, and each prediction of the theory should be found to appear as a phenomenon. While this is the case with quantum theory as applied to the laboratory physics of fundamental particles, it is grossly

---

[8]   Hartle (1991).
[9]   Isham and Butterfield (1998).

not the case when applied to the world as a whole — but, typically, the failure is of the 'mad genius' variety. Whereas a merely bad theory just produces wrong calculations, *quantum theory correctly explains all the experimental results, but also contains accounts of phenomena that could not possibly exist:* it contains vast numbers of mathematical propositions that cannot possibly have corresponding natural language propositions. These are the notorious 'superpositions of macroscopic states' whose natural language counterparts, if they existed, would be things like 'this cat is in a superposition of being alive and dead'. One could say that quantum mechanics is incomplete, in that it lacks the physical laws that are needed in order to rule out these surplus propositions that do not occur in experience.

If we expect science to give a complete picture of the whole of the experienced universe, then this is a severe blow. But it could equally, I propose, be treated as the opportunity for which human culture has been waiting. Two courses open before us, with the choice likely to emerge with a few decades. On one course, physics may be completed as experimentation and theory-building proceed. New laws, which may be among those already proposed by some theorists, may be verified which determine at each context just what range of propositions is eligible to enter a history. In that case, physics will form a self consistent unit. It will continue to be indeterminate, leaving it open to chance which propositions out of the allowable range are fulfilled, and it could be that other discourses alongside physics can operate within the area where they have always held sway, but now separated from the discourse of science. But there will be little interaction between physics and mysticism. In such a situation it would remain an uphill struggle to bridge the gap between the mechanistic view of physics and the needs of planetary culture.

On the other course, *the completion of physics will come through connecting it with what is now outside physics.* Considerations of meaning and value will be seen to determine the allowed range of propositions at each moment. There will be an open interface between physics and the rest of human culture, with physics displaying the ground within which embodiment takes place, including the openness of quantum logic, and the wider culture and the diversity of societies feeding back to physics the contexts within which physics assigns probabilities to histories of manifestations.

On this second scenario, just as the logic of classical physics becomes widened to quantum logic so as to interface with society, so the Aristotelian logic of the dominant society becomes generalised to bilogic or a similar context dependent logic in order to interface with other ways of knowing. The progressive enlargement of logic, and its ultimate dissolution in mysticism, provides the framework in which many ways of knowing can not only coexist, but mutually enrich each other.

The Metaphysics of Participation

The picture that emerges from the generalised histories approach is of the universe as a great conversation, rather than as a machine. As do all beings, I find myself at each moment standing within a web of connections — the connections of physical causation and physical sensation embodied in the dynamics of quantum theory, and also perhaps the more subtle connections of entanglement implied by the structure of quantum theory. I act, and am acted upon, within this web, and as this happens I embody the meanings and purposes that make up who I am, expressed through rational decisions, through the non-rational patterns of being that contribute to the relational side of my mind, and through the silent processes of my body. My enaction establishes a proposition at this here-and-now, within the flexible constraints of the probabilities of the histories approach, and a history emerges from all our propositions, a fragment of which is then reflected back to me, and I see the world. The web ripples out in space, and to the future, and to the past. Its connections are so rich that in many places the probabilities approach certainty, and we can speak of the objective world, and of Newtonian physics. Every action is an *inter*-action, so that I can never observe without at the same time creating.

How far is this picture recognisable in our experience? In my own experience, this is variable. There are indeed times, after meditation out of doors, or during a long walk when the rhythm of the paces has lulled my internal chatter, when I feel so much one with my surroundings that observation turns into communion, and I know that this is what my body-mind was evolved for. But much more often I am withdrawn from the world, receiving but not giving, acting but not being myself transformed. There is an estrangement from the world, and from my body, that seems to prevent a reciprocity in my relation with other beings, just as it reinforces the propositional side of my mind and weakens the relational side.

All the evidence suggests that the historical process in the West has been one of progressive enculturation into the second of these two ways of being and knowing — the estranged rather than the connected. The fact that there is an approach to physics (I would say, the most effective approach to physics) which affirms the connected way of being suggests strongly that our modern estrangement is not simply a reflection of the objective reality of the universe; it is a particular cultural phenomenon. I can myself vouch for the fact that this phenomenon can be reversed; we can learn to open up our connections to the world, to become more aware, more often, of the creative reciprocity that under-lies our relationships with all beings.

There is a fascinating paradox about this that takes us to the heart of context-dependent logic (or, as it is often called, both/and thinking) concerning 'reality'. When I open myself to the presence of a tree, feel the weight of its boughs in my own limbs, let the smell of its bark soak into my feelings so that 'I' drift into the background, then that tree becomes more and more real to me. It moves from being a fact to being a presence whose reality cannot for a second be doubted, as it engages more and more of the relational subsystem of my being. Yet the philoso-phy of 'realism' would have it that reality consists only in the extent to which knowledge becomes objective — that is, unengaged, separated from feelings. It seems that 'reality' subsists both in the greatest detach-ment from feelings, and the greatest immersion in feelings!

Context is the key to this riddle, and it is worth looking at the way in which these two poles are indeed 'reality', but are context dependent. One context is that of physical science, which discloses, and allows peo-ple to engage with, an amazing cosmos that constantly outstrips all we could conceive of. The dynamic of the cosmos that has brought the earth and ourselves into being has all the marks of 'reality'; to suggest that this sort of knowing is just a human invention, as some adherents of a relativist view of truth might say, seems almost blasphemous. It is a vital part of the conversation that is the world, a conversation between the human species and the entire cosmos. As we now call the being of the unified planet *Gaia*, so we might call the being of the unified cosmos *Ouranos*, and rejoice that we can share in our species' conversation with him. But there are other contexts such as my presence with the tree, or the context of a group of likeminded people silent in a forest glade. For these there is a different truth and a different, but no less 'real', reality.

So there is a choice that is implicitly before our societies, though largely unrecognised. We can continue to regard our estranged condi-tion as an inevitable part of a universe whose only reality is the objectiv-ity of a depersonalised consensus, bound to a propositional way of knowing that devalues the relational side of our being; or we can enlarge our world to include, along with this, the more vibrant reality of the relationships on which the whole of our being ultimately rests. The previous authors have charted the way into this second option, and I hope now to have shown how, in taking this path, we fulfil and redeem — not abandon — the ongoing process of science which has both enriched and imperilled our society.

Lyn D. Andrews

Ways of Knowing and the Quest for Integration

I have started this essay three times now, with each new attempt being aimed at reflecting the current status of my thinking, possibly my Being. However, this time things will different; I wish to take a more creative, unconventional approach, as I used to do when I wrote stories, and then later when I wrote *The Cosmic Connection and the Theory of Universal Fulfilment.*[1]

The writing of this essay is, in fact, the unfolding of my life's latest chapter, and like all good stories it contains within it seeds of the past and the future as well as clues as to my character. For a short time, therefore, I'd like to share with you some of my most interesting experiences and explore the way I came to understand and accept them. By doing this it will be found that among the different ways of knowing there is even a way of knowing which can help us manifest our destiny!

To do this I plan to take the advice of Jack Bickham in the *Writer's Digest Handbook of Novel Writing* concerning 'scene' and 'sequel'.[2] The scenes will be about what happened in the past as well as what is happening in the present and will be written in the first person. The sequels, on the other hand, will be reflections and written in the third person.

In this way the scenes and sequels, in other words, the action and reflection, will be seen as part of an integrated whole; a process, which I propose, represents the evolution of the soul. I am also hoping to address the following points, briefly. The concept of intuitive ways of knowing in contrast to more dominant ways of knowing, both/and thinking as it pertains to ways of knowing or what I have previously

---

[1]   Andrews (1999).
[2]   Bickham (1992), pp. 113–121.

called symbolic sight and the issue of hierarchy as it pertains to emotional, intellectual and spiritual growth.

Scene 1, Part 1

It is 1994 and I am 35 years old. I am stationed in the unfinished entrance hall of a very large house in the midst of renovation. Since the study is still to be finished our first personal computer has been set up in the entrance hall instead, with views facing out onto surrounding hills and dissolving into the mountain ranges beyond. Unexpectedly, the odour of pet mice is pungent. The teenage children of the previous owners had kept them. But this is not all. There is another unexpected feeling surrounding myself and the computer, one that can only be described as attraction, much as new-found friends, even lovers, experience. However, I have to admit that before that time I had never befriended a machine let alone fallen in love with one, particularly one that I knew nothing about!

After our initial introduction, however, we are inseparable, and the time we spend together, day and night, is filled with love and creativity that is sometimes spontaneous and sometimes fraught with pain and anguish.

Skipping ahead two years, I have now produced two very amateur, unpublished novels and am eager to produce a third. The next one is to be called *The God Factor* and by means of appropriate characters, a young, female geneticist and a young Catholic Priest, I plan to explore the relationship between science and religion.

Unfortunately, however, I am not up to the task. While I am adequately qualified to deal with the science side of things I have little experience of religion and no experience of overseas travel. Therefore my arguments, as they concern religion, and my descriptions of the overseas locations involved, are bound to be naive and shallow. So I give the idea up and go into mourning.

However, there is an unexpected turn of events. The film *Lawrence of Arabia* is due to come on television, and since I have always wanted to watch it — ever since I was a young child and recall my mother enthusing about the sound track — I am excited and expectant and determined not to miss such a golden opportunity.

Lawrence proves to be another great attraction and I suddenly find myself enchanted by his brilliance, his uniqueness and the mystery surrounding his character. I also find myself compelled to borrow his book, the magnificent *Seven Pillars of Wisdom*[3] from the local library after my father-in-law informs me of its existence during a late night tele-

---

[3]   Lawrence (1962).

phone conversation. They say that all things lead to God; in my case it is all things led to Lawrence!

My enchantment with Lawrence continues and soon blossoms into love. But the love I feel for Lawrence is not specifically romantic, although it has aspects of that. My physical attraction to him is inseparably mixed with empathy and compassion. I seem to understand him in the same way that I am learning to understand myself, and when I say 'learning to understand myself' I am thinking about the role the writing has played in this. Having deliberately placed myself into so many different roles in the stories I have written, I have learnt a great deal about myself, to the extent that it is having an effect on my life. I am beginning to wake up to myself, stand up for myself and challenge the status quo.

Consequently, I begin to analyse his book as well as numerous biographies. I tear his book to shreds both metaphorically and literally, especially the chapter entitled *Myself*, and I psychologically 'assay' him as best I can. I still have a 28000 word essay on floppy disk to prove it! From that sometimes grim and yet emotionally and intellectually satisfying experience I gather that Lawrence's main problem was fear. I also gather it is mine. I also begin praying to God, Jesus in particular, for his deliverance, and saying things, while dancing and/or crying like, 'Please God, just let him be happy'.

By this time I have already started a third book. It is called *Miracle of the Desert: The Untold Story* and it is all about Lawrence. I feature in it too, staring as a gypsy seer who falls in love with him and suffers as he does during and after the First World War. It is during the writing of this that I lapse into trance at the computer and experience what I now believe to be mystical union.

Scene 1, Part 2

Approximately two weeks after the Easter of 1996, and in the newly renovated study, I am working on the novel I am writing about T.E. Lawrence and myself, and as is often the case, I am fascinated to understand the source of my ideas. I am also very much in love with the process of creativity and very much in love with T.E. Lawrence, who is proving to be both the fascinating and tragic star of my story as well as a positive influence in my life.

So on this occasion, and as I have done many times before, I pose the question, 'Where do my ideas come from?' I do this because sometimes, not always, ideas for my stories don't seem to come from me at all, but from somewhere else, somewhere deep and mysterious, and I am incredibly curious about this. Then suddenly, as if in reply to my question, there is a kind of click inside my mind, or a switch of some kind, which alerts me to the fact that something unexpected is about to happen and that my approval is required. I unhesitatingly agree.

The unexpected occurrence turns out to be a mystical trance experience, which unfolds in four distinct and memorable phases described below.

Phase 1

This phase has elements, which as a storyteller, I can easily relate to. I immediately realise that just like the characters in the stories that I write, I am a character in someone else's story. As well as that I understand and accept that I exist in and of the mind of God. But this is not all. In almost the same moment of clarity I realise that in order for one of God's characters, such as myself, to function as a co-creative individual, there has to be something associated with their minds which, paradoxically, separate them from everybody else's mind but not from the mind of God.

Phase 2

This phase of the trance is slightly harder to describe because by this stage my sense of separateness is diminishing and I find myself in a complex web of interconnectivity that has depth and breadth as well as a great deal of meaning.

Phase 3

This phase is easy to describe because at this stage I lose consciousness altogether. So there is actually nothing to describe. My mind goes blank but this does not mean empty. On the contrary, when I begin to regain consciousness at the beginning of Phase 4 my mind is full of the feeling that I have experienced everything, that I am everything!

Phase 4

During this phase of the trance I see and experience myself as a spinning thought of God and as part and parcel of an enormous system of other spinning thoughts of God, where everything is itself and yet interconnected and spinning slowly in a spiral or vortex.

Sequel 1

The experience came as quite a shock and triggered another very eventful phase in my life, a phase which is really only coming to a close now. After regaining consciousness I sat at the computer and shook my head in amazement and said to myself, 'What was that?' Vaguely though, I knew what it was, but the enormity of it was intimidating.

Somewhere deep inside of me, which was, as it turns out, my intuition again, I knew that I had to figure it all out and tell someone about it, someone with power — intellectual power — and someone with strong connections to the academic world. This seemed absurd, egotistical. After all, who was I to tell anybody anything? But by this time, thanks to the creative writing and my deep affection for T.E. Lawrence, I was able to differentiate between love and fear. This meant that I was not going to make the mistake of turning my back on something I loved and there was no doubt in my mind that I loved what I had seen.

While the experience was over in a flash it has taken me years to recover and in many ways I am still in recovery. Those few seconds or whatever length of time it was (I really don't know) changed my life forever. How could it be otherwise, when I was so compellingly drawn to the adventure, and then, having set out upon it, temporarily lost consciousness and bore witness to what I now regard as four different structures or aspects of consciousness, possibly reality, possibly God? I changed from being someone who not so much denied God, as ignored Him, into someone who had little or no doubt that there actually is a spiritual reality. Not only that, but a spiritual reality which is composed of structures of consciousness which are pre-personal, personal and transpersonal. In a moment of clarity I seem to have taken on board a whole new world-view, but was, at that time, unable to articulate it fully. However, since I was keen to do so, I made it a priority.

However, when I first tried it out on someone they scoffed at me, got angry, and I quickly realised that an experience such as mine is very difficult to explain. But that didn't deter me. I simply couldn't contain it. The love, the bliss which followed, which I felt for nearly three weeks after the trance, put paid to any quaint ideas about me trying to cover it up or forget about it, and then when I started hearing the voice, the game was up. I was a mystic and that was all there was to it, and not only that, I was going to tell someone who might be able to do something about it. But how, was the question. After all, at the time I was an unemployed, former secondary science teacher/homemaker whose contact with the academic and religious world was non-existent. In addition, I was only beginning to come out of my shell, emotionally and intellectually speaking. The idea of presenting something altogether new to people more qualified than I was, was daunting, even terrifying. But again, something deep inside me said, 'Just keep going' which basically meant, stop fussing and be guided by my heart and follow up with my head.

During the following period I was busy in an intellectual sense. Desperate to turn my personal experience into something communicable and relevant to others I endeavoured to become better informed and this meant catching up academically. However, due to my current academic status, as well as my geographical location, I knew I was at a disadvantage. Therefore to compensate for my various shortcomings I bought and read good quality popular science and psychology.

In terms of doing research, there was no doubt that I was attracted to theoretical physics first, psychology second and mysticism third. I was also interested in the way my experience related to interpersonal communication and education. I really didn't need to read much about mysticism because I was already, post-trance, a mystic! What could another mystic tell me that I didn't already know? My main purpose in reading the mystics was to assure myself that I wasn't going mad. However, the chances of me becoming a theoretical physicist and/or a psychologist were slim. I therefore started a (sometimes desperate) search for a world-class theoretical physicist who was willing to listen to me and help me.

The realisation that I needed help, including the kind of person who should help me, could be regarded as an intuitive way of knowing and such a 'way of knowing' was going to continue to play a leading role in the story. That knowing, which I had become more aware of during my periods of creative writing, I was beginning to trust. I was to gradually realise that, while such a way of knowing does not provide all the answers, it stimulates, possibly initiates the quest, and that the key to success lay in matching up — integrating — various ways of knowing. Even our complaints, however minor, can be a clue to the next phase of evolution. However, in everyday terms, intuition can make us restless and therefore more likely to ask questions and go in search of suitable answers.

It took me three years to make contact with Chris Clarke and by the time I did, I had a bit more of a handle on things, in terms of the physics involved, but not much. My ideas about quantum theory, for instance, were still pretty vague. However, in other respects I had achieved a great deal. The terminology I used was fundamentally my own, and when I found that some of it was already in use in other published material, especially *A Course in Miracles*[4] and *The Urantia Book*,[5] this served to encourage me further. This was purely for the purposes of validation. I had come to realise that very few people were going to accept my ideas at face value, especially those belonging to the scientific community. Therefore, my job, it seemed, was to support my experience with evidence. So instead of writing a novel aimed at integrating science and religion, I was writing a thesis aimed at integrating science and religion

[4]   *Course in Miracles* (1975).
[5]   Urantia Foundation (1955).

using grounded theory as my methodology. In addition, instead of trying to integrate genetics and Catholicism I was trying to integrate theoretical physics, psychology and religion. This means I was actually living the theory at the same time I was writing it! It certainly came as a shock to me, though a highly stimulating and intriguing one, and while I was lacking the usual kind of supervision I was not completing lacking in it since post-trance I could 'hear' the spirit which resided within me, of which I shall speak more later.

The material in *A Course in Miracles* reflected my experience perfectly, especially the sections related to removing judgement, the projection of the ego, the forgiveness of oneself through the forgiveness of another (in my case, T.E. Lawrence), its emphasis on the role of ego and spirit, and its description of the decent of grace and the way revelation can arise from this. As amazing as it now seems, this is exactly what happened to me, thanks to my love of creativity and for T.E. Lawrence. So it isn't surprising that I now take *A Course in Miracles* very seriously and see it as a guide by which self-centeredness and narcissism, both of which I believe arise from fear, can be recognised and possibly transcended. However, I am still undecided as to what I think is really going on. It seems to me that spiritual growth is mixed up with intellectual growth and they are both mixed up with practical experience. I believe that there are particular levels of integrity to be achieved rather than particular levels of consciousness to be reached. This would help dispel the confusion which surrounds a transformative approach to spirituality and a hierarchical one and it brings me to how I came by these ideas in the first place.

The transformation in my life was far reaching. In addition to those already mentioned, post trance there were numerous repercussions. For a start, it promoted my already established need to straighten out the relationships in my life. This restoration program had already been started in relation to my eldest son, and thanks to T.E. Lawrence I had already realised the mistake I had made, which wasn't at all unusual but it was profound. I had, on many occasions confused conditional love with unconditional love and then wondered why it didn't feel so good. But my relationship with T.E. Lawrence helped me understand all that and I soon began to realise that unconditional love is that which allows a person to be who they are and not what you expect them to be. This, of course, was difficult to put into practice, especially in the case of a young child who was in need of discipline. I found it difficult to differentiate between being a parent whose role it was to set conditions and administer discipline and being a parent who loved their child unconditionally. In the end I established an intellectual and emotional truce with myself. Along with some reasonable boundary setting, I allowed my eldest son to be himself so long as it made him happy. But when it didn't, that is when he was obviously angry or upset or out of control, then I knew something had gone wrong and I stepped in. This meant taking a great deal of notice of him and listening. In this way I could understand, more readily, the reasons for his unhappiness if and when it occurred. Consequently, I now place a great deal of emphasis on setting reasonable boundaries, being present and exercising empathy, which includes listening (especially for complaints) and communication. This is not to say my relationship with my son is now perfect, but it is a whole lot better than it was and we're both happier!

Gradually other repercussions made themselves known and during the past seven years I have experienced many changes in my life and I still am. As soon as one issue was addressed the next one, large or small, showed up in a kind of spiral until this process culminated in the break up of my marriage. This, it seemed, was a climactic chapter in 'getting real' with myself and with others, including God, although as you can imagine, such a series of transformations has not been easy. All in all there were quite a few heartbreaks, set backs, deaths and resurrections.

In the mean time I learnt a great deal. For instance I found that I understood things post trance that I didn't have a clue about pre-trance. While I didn't understand everything, I understood a lot intuitively, as previously indicated. Initially, I accounted for this by accepting that something had happened to me during trance which had prepared me for my purpose. I also began wondering if the Holy Spirit could endow us with spiritual gifts as described by St Paul. If this was the case, then I had been endowed with what I called "symbolic sight". This meant I could see the meaning of things in a way that very few others could. However, I am not so sure about this interpretation now. Other mystics may have experienced the same thing. For instance, the mystic Ibn-'Arabi wrote, 'The natural order may thus be regarded [at once] as [many] forms reflected in a single mirror or as a single form reflected in many mirrors ...' Clarke, who quotes this[6] also informs us that the mirror analogy is echoed by other Christian writers such as Hildegard of Bingen and goes on to say that the mirror image is not meant to imply that one order of reality is higher than another but that the idea of synthesis is important. He says, 'such a synthesis is unimaginable intellectually, but can be encompassed through mystical love', which is, as it turns out, what I think happened to me while I was writing the story about T.E. Lawrence in 1996.

---

[6]   Clarke (1996).

Consequently, I now think that this kind of synthesis, which I initially called symbolic sight, is what Gebser[7] calls, 'integral consciousness' and Chris Clarke[8] calls the logic of both/and. In accord with this is the integration of personality with spirit, which will be described in due course.

In 1996 or 1997 (can't recall which year now) I channelled this line from my own higher consciousness. After the foregoing discussion it makes more sense than ever!

> Internalise then transpose; the universe plays tricks on us. Those that don't like riddles need not apply.

This means that the mind can carve up the universe for its own purposes and if we're not careful, we can believe our own creations to be the truth.

Another contributor to this volume, Isabel Clarke, discusses in more detail than me another way of seeing things. The Interacting Cognitive Subsystem Model (ICSM) presents the mind as being composed of several subsystems, two of which are intimately related to each other. The relational subsystem is thought to be involved with the instinctive reactions of the body, the emotions and emotional meaning and our relationship to the whole, while the propositional subsystem is thought to be involved with the processing of information and verbal coding.

By utilising this particular model to explain what happened to me, I now suspect that when I began writing in 1994 I was utilising the relational subsystem more than previously. I further suspect that the emotional side of my nature is more intimately related to the intellectual side than I previously understood.

However, I now consider that the situation is even more complex than the ICSM implies, as we shall shortly see. In fact, and as previously indicated, I now realise that the ICSM is really just another way the mind can represent itself, and like a snowflake, each representation is unique. However, in order to support this conclusion, it will helpful to explore the mystical trance experience a little more closely.

Immediately after the trance and as previously indicated, I began to record the experience in a meaningful way. One of the first things I did was to describe and name the phases I saw and experienced in trance according to the way I had understood them. Some phases have more than one name, which I know is confusing, but together they are meant to convey both their structural and functional aspects.

The structure I saw and experienced during phase 1 I called personality and regarded it as the means by which we can function as individu-

---

[7]   Gebser (1953).
[8]   His chapter in this volume, pp. 139–158.

als and yet be intimately related to the whole of creation at the same time including mind, matter and spirit.

The structure I saw and experienced during phase 2 I called mind and regarded it as being personal by virtue of the presence of personality as well as shared in the sense that we are all a part of it.

The structure I experienced but didn't actually see during phase 3, I called a variety of names including spirit, creative spirit and the original source of pure possibility. I regarded it as that aspect of consciousness which is the source of all the other aspects of consciousness, as well as being an ever present source of love and creativity.

The structure I saw and experienced during phase 4 I called a variety of names too, including the oversoul of creation, Evolutionary Deity and 'The Actualising Unity of Deity'[9] and regarded it as an integrated system of matter, mind and spirit which is conscious and evolving.

In other words, during that once off experience, I experienced most, if not all, of the experiences which mystics of other traditions have described. These ranged from the dual to the non-dual, and from the personal to the impersonal to the pre-personal. From this I concluded that it is possible to apprehend them all once and that each of them is a part of a magnificent system utilised by God for the purposes of creation.

Despite this terminology above being largely my own invention, and my use of a variety of names complex and confusing, I continued to incorporate them into my developing thesis. Interestingly and unexpectedly, I eventually came across *The Urantia Book*[10] which has an account of the nature of consciousness and reality which uses similar terminology and fits well with my experience.

Like me, *The Urantia Book* distinguishes four aspects of consciousness, which it calls personality, mind, spirit and soul. Like me it distinguishes between personality and mind, where personality, according to *The Urantia Book*, is also that aspect of consciousness which integrates the rest: in other words, the integrated self. Also like me, *The Urantia Book* distinguishes between two phase of spirit. One of these it calls the 'thought adjuster', which indwells the mind of a human being and claims its origin is in the Father, or the First Person of the Holy Trinity. However, in contrast to *The Urantia Book* I prefer to call it pre-personal spirit. The other phase of spirit, and which I prefer to call creative spirit, is said to have its origins in the Holy Trinity.

If all this is to be taken seriously, then the big picture is shaping up to be extremely complex. The nature of consciousness, possibly reality, at

[9]　　Andrews (1999).
[10]　Urantia Foundation (1955).

least in respect to human beings, is also shaping up to be a combined human-God enterprise. So at this point I would like to briefly refer back to the ICSM again.

As previously indicated the ICSM helps explain the various types of processing going on in the brain. There's the relational side of things and the propositional side of things but their exact mode of function remains unclear. However, it is clear that they are intimately related and that their smooth, integrated functioning is important to the health and welfare of the individual; the implication being that a lack of integration could result in the over emphasis of one mode with respect to the other and therefore a degree of inner and possibly outer unrest.

Before 1996, if I had have been studying psychology, I would have accepted a model like this without question. In addition, I may or may not have considered others. However, these days I am inclined to think differently. Instead, I am proposing that the mind, in terms of its thinking and feeling capabilities, is structurally and functionally related to both phases of spirit as well as to personality as well as to energy/matter, and that the conscious realisation of this is an important step in our evolution.

In other words, I am grappling with the idea that what I saw and experienced in trance was not a creation of mind or a model, but the real thing. Ferrer[11] would say, 'things as they really are'. I mean to say that there is a pre-given system of matter (energy) mind and spirit which is utilised both personally and universally in the co-creation of our unique and immortal souls as well as Evolutionary Deity. Ferrer would call it 'spontaneous cocreative participation in an indeterminate spiritual power or Mystery'.

I am therefore claiming, that by being able to distinguish between the creations arising from the system and the system itself, I have moved a step deeper into Ferrer's 'Mystery of being in which we creatively participate'. In doing so, perennialism and participation can be understood to be intimately related. Universalism and pluralism would be mates.

Scene 2

It is still 1996 but later in the year and I am up the other end of our house doing the vacuuming. Suddenly it happens again; I hear the voice, or rather, I receive it. I am not afraid of it, merely annoyed. The thing, whatever it is, just won't leave me alone; it's like having to answer the telephone all the time. I can't get the housework done! Thank God I'm not a secretary. I am angry and tearful and flop onto my eldest son's bed and howl.

---

[11] See Ferrer (2002), and his chapter in this volume, pp. 107–128.

This tantrum seems to do the trick because the next time the voice, or whatever it is, seems to have learnt some manners. It remains in contact but more gently and when it wants to communicate with me it makes my left earlobe tingle and makes me sleepy. I usually respond accordingly and have a short, ten-minute nap on the couch and when I wake up I dream a highly symbolic dream and/or I receive the message thereby establishing a more agreeable working relationship.

One such dream is particularly informative and occurs when I have company. Our new, young bookkeeper is in the study and she is doing the job I used to do but have outsourced due to the pressing nature of other interests, namely my research into consciousness and the nature of reality! This move is hardly surprising because I was never cut out to be a bookkeeper. I only did the job because I felt I had too.

As usual, the experience starts the same way. My left earlobe tingles and I suddenly feel sleepy, and so I cooperatively lie on the family room couch to prepare to go to sleep. I don't take long to drop off, only a few minutes, and before long, I am waking up again, and in keeping with the usual sequence of events, my waking up is accompanied by the feeling that I should pay close attention. I do this and bear witness to, what I later regard, as a symbolic representation of creation.

It begins with an explosion of white light - a white out - which immediately clears to reveal the appearance of a mysterious black dot and the white out clears. However, this mysterious black dot immediately takes on a whole new shape turning into a short, thin black line which, itself, suddenly opens up into a circle that looks like an unimpressive elastic band. While all this is very interesting, it has not prepared me for what happens next. Suddenly the circle changes dramatically and becomes a pair of labia ripe for fertilisation! Immediately following that, a part of the circle, now labia, emerges from itself and then doubles back in kind of integrative loop which I understand represents the self-fertilisation of the universe and a process which is orgasmic! The impact is so profound that I wake up with a self-conscious start and leap off the couch. I also make a beeline to the kitchen table, grab some scrap paper and begin scribbling down my impressions. At the same time I channel two lines, which later, help put the whole thing into perspective. By 'channel' I mean, I receive information from mental press which is nothing like the way I usually think. Its just like someone is talking to me from the inside.

Sequel 2

So what does all this mean? As far as I am concerned it means a variety of inter-related things.

- Creation is deliberate
- Creation is participatory
- I was in communication with something which was far better informed than I was and which seemed able to communicate with me in a variety of ways.

So let's address these one at a time.

Creation is deliberate

First impressions of the dream suggested that the explosion of white light was symbolic of the Big Bang and the raising of human consciousness or the so-called fall. The implication was that the Big Bang was related to the emergence of matter from spirit and the self-fertilisation of the universe was related to the ability of human beings to participate with spirit in creation. This impressed me greatly, especially since the mystical trance experience had implied much the same thing.

In a sense, it is in this way that God reproduces Himself over and over again, thus making his creations co-creative. (No wonder the experience was orgasmic!) Not that I knew it then, but this raising of consciousness I later attributed to the gift of personality, which is, according to *The Urantia Book*,[12] an actual gift from God, the Father, as previously mentioned, and an aspect of consciousness which distinguishes us from our purely evolutionary origins.

Creation is participatory

This has largely been covered in the previous section but in addition I would like to say that the dream implied a vital significance to the self-fertilisation of the universe in terms of human participation. Like the trance had before it, it highlighted the importance of the gift of personality as the means by which human beings are inherently co-creative.

I was in communication with something which was far better informed than I was and which seemed able to communicate with me in a variety of ways.

These means of communication had already included a trance, dreams and mental presses (direct contact) as well as my emotions, which incidentally were, by far, the most difficult to figure out. So naturally I was curious and even a bit dubious and considered mental illness. But luckily for me, I knew a retired psychiatrist and went to see him. However, I deliberately skipped telling him the bit about the trance and the voice and came away with some books on psychiatry including a copy of the

[12]   Urantia Foundation (1955).

Diagnostic Statistical Manual of Mental Disorders[13] better known as the *DSM* in order to read up on psychosis.

While my experiences had been challenging, and continued to be so, I found nothing that pointed to severe mental illness. I did, however, own up to my need to change in relation to myself and to others and recognised that I was in the process of doing that anyway. Overall, I considered that I was making the necessary adjustments in my own time and in my own way. So while I gave myself a relatively clean bill of mental health I also recognised how close I had come to tipping over the edge. That's when I re-read Somerset Maugham's classic novel *The Razor's Edge*[14] and appreciated it more than ever. I also read Hillman's *The Soul's Code: In Search of Character and Calling.*[15]

James Hillman fortuitously introduced me to the idea of the daimon, other names for which are soul companion and genius. I also read M Scott Peck[16] and thought how interesting it was that two renowned mental health specialists didn't see eye to eye. To me they both had something valuable to say and I went about integrating their different points of view.

I accepted some of Hillman's ideas about the daimon, especially as it relates to the call and, due to M Scott Peck, reintroduced the idea of the ego as being the big block to self-realisation, where ego here means an attitude of mind which limits self acceptance and authentic self expression. This meant I could account for the voice as being related to the daimon, which I now believe is the way that pre-personal spirit can communicate with us directly, and I could account for the ego being that part of the mind which is fixed, inhibited or distorted. However, I understood the structures which I had named personality and soul in the same way as previously; personality being that structure of consciousness which is a gift from God (as recognised during trance and the dream) and relatively co-creative, and soul being the evolutionary accomplishment of the combined activities of personality, pre-personal spirit and lived experience which involves the temporary relationship with mind and matter as we know it.

Thus, from both personal experience and published material, I had a developed picture of consciousness which included structures such as personality, mind, pre-personal spirit, soul and ego, where the ego is a product of mind, a pattern of thinking (in Buddhism and Hinduism an illusion) and not really a structure at all. I had also realised that our ability to differentiate between the call of pre-personal spirit and the ego is

[13]   DSM-III-R (1987).
[14]   Maugham (1963).
[15]   Hillman (1996).
[16]   Peck (1978).

largely the same as having the ability to differentiate between love and fear. I had therefore made the same journey intellectually that I had emotionally!

The implications of this are enormous. It means that we are two minded creatures! One mind, the material or mortal mind, is related to our everyday thoughts and feelings as well as to our basic instincts as they pertain to the body and our means of survival. We also have another mind, as in pre-personal spirit, which is related to the whole, as well as to love and creativity. We are, in a sense, between worlds, (Heaven and Earth) as well as being 'two brained', as *The Urantia Book* says. More simply, however, I now understand the nature of consciousness, possibly reality, to be a complex combination of human and divine structures which are intimately related to each other as well as to material creation.

Scene 3

It is now 1999 and I have read a great deal and have begun to integrate my ideas into a comprehensive picture. I have also written *The Cosmic Connection and the Theory of Universal Fulfillment* [17] and published it on the Internet as the voice had unexpectedly recommended. But I am still confused about quantum theory and I am still very emotionally and intellectually lonely. I long for someone sensitive and well informed to talk to and confide in. So, in one last desperate bid I contact the Theosophical Society and they pass me onto someone who might be able to help. As it turns out they can (so I don't know why I didn't contact the Theosophical Society before), and I am given the website address of The Scientific and Medical Network and the name of a book to read, *The Spirit of Science*. [18]

I check out the website immediately, join up immediately and likewise order the book immediately. There is no doubt in my mind they are the people I have been searching for.

But before I go too far too quickly I check out the personnel. I read a few things they have written. David seems okay, open minded, and I like the idea of him being a writer and an editor. Chris seems okay too and I especially like the idea of him being a spiritually inclined, theoretical physicist. What a stroke of luck! I make plans to write to them both.

Although it is a dream come true I am afraid to write. Before, when I had written to others, there was only a slim chance of receiving a response, but this time it is different, I have just joined their group, so they are bound to write back, and when they do, I am not sure how to

---

[17] Andrews (1999).
[18] Lorimer ed. (1998).

handle it. I am hugely intimidated. But I find the courage to make contact and they both prove to be exactly what I was looking for. Thank Goodness!

Sequel 3

From the beginning I had a hunch about quantum theory but I was not nearly well enough informed to know if I was on the right track or not. I also had the feeling that what I had to offer might help fill the gaps in the current knowledge.

When I first wrote to Chris I had the idea that quantum theory was about what might or might not manifest, but what I could not figure out was how the outcomes could be accounted for, and as it turned out, neither could anybody else; quantum theory was actually more about possibility than actuality. This didn't deter me though, instead it inspired me, and I began to wonder if, indeed, I did have the answer! However, before I found the courage to present a proposal I had to get to know both quantum theory and Chris a little bit better.

After quite some time it dawned on me (it had already dawned on Chris) that the outcomes of quantum theory were context dependent. In other words, not only was creation pregnant with possibility but each particular situation was different and could result in a different outcome. In other words, every creation is unique! Given time, patterns arose and in consequence statistics could be gathered, but still, this did not explain exactly what was going on beneath the surface. In fact, once I understood this, I began to explore the so-called 'central mystery of quantum theory'[19] all over again.

To begin with, I reasoned that the context setting had to do with us, that is, *personal mind:* a structure of consciousness that is paradoxically associated with personality as well as the whole as seen and experienced in trance. I also reasoned that personal mind functions in conjunction with *pre-personal spirit,* something I had experienced as distinct mental presses or inner communications post-trance. I distinguished this from *creative spirit* which functions in conjunction with personal mind and pre-personal spirit — since I had seen and experienced this in trance as well! This means that there are two aspects of creativity going on at the same time, an inner aspect of creativity related to pre-personal spirit and an outer aspect of creativity related to creative spirit. I also reasoned that creativity involves doing something to/with our thoughts which can be totally unexpected and in contrast to our more familiar, sometimes habitual way of thinking.

---

[19]   Gribbin (1995).

However, the actual process is, as you might imagine, not particularly straight forward because the outcome(s) of such creative events are unpredictable and yet they make perfect sense. Thus, it seems that while we contribute our thoughts and feelings to the creative process using the gift of personality, it is spirit which actually closes the deal. However, this hypothesis does not answer all the questions. While it might be fine while we're talking about creativity in respect to our thoughts and feelings, what about matter and its associated physical processes? How do they fit into the picture?

As it happens, this question ties in with the confusion surrounding the idea of the state reduction in quantum theory. If I am right, and while creative spirit is doing its thing, there is little chance of anyone saying with certainty what will arise or manifest. It is not merely that the outcome between alternatives is uncertain, as in tossing a die, but that even the range of possibilities is uncertain.[20] Another way of putting it might be to say that the whole and the part are intimately related and that the whole and the part include matter, mind and spirit.

In respect to wave patterns, therefore, or indeed any other manifestation, it is not only the emergent pattern or manifestation which is of interest, but the fact that the outcome alters when the context changes; the implication being that reality has a relationship to the past, the present and the future as well!

This being the case, it would be very difficult to differentiate between the integrated activity of pre-personal and creative spirit, the more pattern-loving and predictable activity of personal and universal mind and what we assume to be normal physical processes without the aid of revelation.

Scene 4

It is 2004 and I am 45 years old. Ten years have elapsed since I first began writing and nearly eight since I enjoyed mystical union. In an effort to understand and articulate my experiences I have written a great many essays, articles and emails, of which three articles have been published. I have also attended three overseas conferences and given a presentation at one. Not much in comparison to most of the other contributors, but for me, the unexpected journey has been surprisingly successful. However, after all the changes which have taken place life remains a challenge, but it is, at least, authentic, and I have, at last, the opportunity to tell my story.

---

[20]   Clarke (2002).

Sequel 4

As it turns out it is really very simple. The source of my knowing is not really what I thought it was at all but a reflection of what I am and what I am trying to be. *Knowing, therefore, is related to creation.* By virtue of my pre-personal spirit I am in touch with the inner me, and by virtue of that same source I am in touch with the whole of creation. However, there is a catch. To know this is to be aligned, not with the ego but with spirit, pre-personal and creative, and therefore to be one with God. This might be what Jesus meant by the Kingdom too, and what it means to say that to enter the Kingdom means having a new relationship with the world.[21] For me, at least, and in terms of everyday life, it means being authentic, real, and not necessarily what I'd like to be but what I was cut out to be. It means being receptive to spirit. Therefore freedom is not about doing our own thing, or towing anyone else's line, but choosing to do God's thing, which, as it turns out, will suit us, and everyone else, best. Thus, the education of the future will not entail more information transfer or act as the springboard to a particular career. It will be about assisting spirit to realise a person's true and ongoing potential. It will be about the evolution of the soul, personal and universal. Why? Because spirit is the business of re-creating itself in as many different ways as it possibly can and there's no stopping it.

In closing I just want to add that I think Ferrer[22] is right when he says:

> By the emancipatory power of spiritual truths I mean their capability to free individuals, communities, and cultures from gross and subtle forms of narcissism, egocentrism, and self-centeredness.

However, I would also like to add that in terms of the system of matter, mind and spirit which I present, emancipation implies a personality which is aligned with spirit. When a personality is aligned with spirit it no longer suffers from any 'gross and subtle forms of narcissism, egocentrism, and self-centeredness'.[23] It is free! This means, of course, that the Mystery, with which a human being can then fully and consciously engage, is their own, as well as God's.

---

[21]   Clarke (2002).
[22]   This volume, p. 124.
[23]   *Ibid.*

The Nature of the Spiritual Path

The previous chapters, while using at times different conceptualisations and models, have presented a common theme: humanity can, and — if it is to flourish and even survive — *must* restore its personal, cultural and cosmic connections with the subjugated knowings of the mystical/relational/symmetrical/affective side of being. The theories have been presented, the stories told, and now 'what must I do to be saved?'[1] What are the actions and ways of life that can turn this into practice? Many of the previous chapters (especially those of Bomford and Ferrer) have touched on mystical knowings that are rooted in rich faith traditions and highly developed practices which in former centuries were dominant ways of knowing. In today's more pluralistic world, and within the pluralistic picture that this volume has painted, we need a wider perspective of how we can carry forward our spiritual traditions to meet our urgent need. The transformation of our concepts that has been portrayed will demand a transformation of our practice. This last section offers three contributions to this, which all also continue to extend our conceptual understanding of our situation.

For many of us in the West, the problem is that the mainstream Christian tradition seems bankrupt, for the reasons described by Watt above, while the traditional forms of Hindu spirituality seem hard to connect with the Western empirical approach to science (though others would contest this). In this context, Buddhist thought has contributed much, particularly within psychology, but as this is well covered elsewhere I have chosen not to repeat that work here. Particular interest, however, attaches to Middle Eastern mysticism, which can be conceptually as well as geographically intermediate between West and East. In his chapter below on Middle Eastern spirituality, Neil Douglas-Klotz

---

[1]   Acts 16.30, *New International Version of Bible*.

makes available for us ways of knowing derived from Hebrew and Islamic traditions. This reveals both the differences and the similarities between the paths of these text-based traditions and the spontaneous mystical experiences explored in other chapters. From the traditions he formulates a spiritual practice termed a 'hermeneutics of indeterminacy' whose characteristics strongly parallel those introduced above by June Boyce-Tillman in describing subjugated ways of knowing. His account, which deserves careful reading, unfolds in detail a way in which all the monotheisms can be 'rehabilitated' into the present day. From here he moves to formulate a 'mysticism of everyday life', which he then develops by comparing with similar ideas found in the writings of the humanistic psychologists Abraham Maslow and Wilhelm Reich. This in turn makes links with previous chapters by Elam and Holt commenting on the assumptions of the psychiatric system, showing the centrality of these Middle Eastern insights to an understanding of our theme.

This revisioning of everyday life is found also in David Abram, who has also been strongly influenced by non-Western thought — in his case by the thinking of indigenous peoples which links with subjugated ways of knowing. His chapter introduces a radical new dimension into the whole discussion in arguing powerfully that we can make sense of the confusion of different landscapes, different worlds, which now confronts us, by finding a rich and fertile common ground from which they all spring. He writes:

> what a boon it would be to discover a specific scape that lies at the heart of all these others. For if there is such a secret world among all these — if there is a specific realm that provides the soil and support for all these others — then that primordial zone would somehow contain, hidden within its fertile topology, a gateway onto each of these other landscapes.

And he proceeds to find this ground as "none other than the sensorial terrain of tastes and textures and ever-shifting shadows in which we find ourselves bodily immersed." From here he unfolds the liberating consequences for our lives of this vision of the core ground of all ways of knowing. His account also brings home our vital dependence on all the beings with whom we humans share this planet, a vital corrective to the human-centred vision of many spiritualities.

I have given the last chapter to Anne Primavesi whose conception of 'Ecological Awareness' brings together into a practical path the most important strands in this book. Ecology is central to her vision. In links into spirituality, in the way already vividly described by Abram; it is a scientific discipline that is crying out to be taken far more seriously as a central part of our teaching and support for science; in the form of ecopsychology[2] it also unites the psychological perspectives with Abram's spirituality; and in its recovery of the thinking of Bateson it opens new avenues in our philosophical thought about the nature of the world. Writing from the Christian tradition, her chapter shows in particular how this also can be recovered and freed from the encrustations that Watt has described.

[2]   See Fisher (2002).

Neil Douglas-Klotz

Ordinary and Extra-ordinary Ways of Knowing in Islamic Mysticism

---

Introduction

Ancient ways of knowing have been dismissed by some scholars as evidence that our ancestors lived in a sort of 'magical' universe, one which has no application to post-modern ways of living informed by post-Enlightenment scientific thought. More sustainable knowings supplanted the ancient mind, according to this view, simply because they were more adaptable to a 'truer' understanding of the nature of reality. However, given the state of the world that these 'more adaptable' ways of knowing have bequeathed to us, it may behove us to take another look at traditions of holistic thought that survived the Enlightenment and which prove very adaptable for living in a world of diverse worldviews.

The Middle Eastern Jewish and Islamic traditions have in common their rootedness in sacred texts, written in the closely related Semitic languages of Hebrew and Arabic. They have evolved a highly articulated mystical 'hermeneutics' (that is, a spiritual practice of interpretation) which reveals a rich and open-ended way of knowing. I first investigate the Islamic version, called *ta'wil*, and compare it to the work of social science researchers Reason and Rowan[1] on ways of construct-

---

[1]   Reason and Rowan (1981).

ing new paradigm inquiry strategies.[2] This leads to my formulation of a 'hermeneutics of indeterminacy'[3] as a way of reading Biblical and Quranic texts in order to understand non-Western ways of knowing as well as the Semitic language spiritual experiences that these texts may embody. In the spirit of *ta'wil*, we will be facilitating a conversation among mystics across the ages.

The chapter next proceeds from the general to the specific and investigates the ways in which various classical Sufi writers attempted to articulate the relation of 'ordinary' to 'non-ordinary' states of consciousness. I compare classical Sufi descriptions of a mystical state (*hal*) and mystical station (*maqam*) with modern and post-modern concerns about a 'mysticism of everyday life'. The experience of a *hal* denotes a state of grace that descends upon a Sufi practitioner, but which is only temporary and facilitates a new 'station' (*maqam*) in life that represents the ability to bring a visionary state into everyday life.[4] Finally, the chapter compares these ideas to similar ones found in the writings of humanistic psychologists Abraham Maslow[5] and Wilhelm Reich.[6] In both classical Sufi terminology and practice, as well as in the theories of Maslow and Reich, one finds the attempt to contextualize 'everyday life' itself within a mystical framework, that is, not only is there a mysticism of everyday life, but everyday life itself is seen in an extra-ordinary way, as a type of mysticism in itself. As Maslow puts it:

> The great lesson from the true mystics, from the Zen monks, and now also from the Humanistic and Transpersonal psychologists — that the sacred is *in* the ordinary, that it is to be found in one's daily life, in one's neighbors, friends, and family, in one's back yard, and that travel may be a *flight* from confronting the sacred — this lesson can easily be lost. To be looking elsewhere for miracles is to me a sure sign of ignorance that *everything* is miraculous.[7]

Re-hearing the Quran: Ta'wil and a Hermeneutics of Indeterminacy

The basis for the *ta'wil* approach to hermeneutics (interpretation) lies in the qualities of the Semitic languages that lead to ambiguity in the meaning of a particular text. Both Jewish and Islamic traditions of mystical hermeneutics point to the importance of individual letters and letter-combinations. The Semitic languages depend upon a root-and-

---

[2] A similar discussion could be considered in relation to the Jewish mystical tradition of *midrash* — for instance, see Douglas-Klotz (1999) — which the Islamic tradition of *ta'wil* parallels in many respects.

[3] Douglas-Klotz (1999, 2000, 2001, 2002).

[4] Nasr (1991), Schimmel (1975), Ernst (1997).

[5] Maslow (1968, 1993).

[6] Reich (1948, 1949).

[7] Maslow (1993), p. 333.

pattern system that allow a text to be rendered literally in several different ways.

This root-and-pattern system, and the interpretive methods that evolved from it, could be compared to the musical system of Indian ragas in which families of notes and scales interlink and 'intermarry' to produce other scales. The closest equivalent in Western music is the free-form improvisations on a theme found in jazz. Like jazz and raga, learning this type of translation-interpretation, seemed to depend as much upon feeling as upon technique, as much upon individual contemplative experience as upon scholarship. Particularly in Kabbalistic and Sufi circles, these techniques were passed on in an oral tradition that included a community of voices, both present and past, upon which subsequent interpretations were built, using the possibilities in the language as well as traditional stories and folklore.

Classical Ismaili and Sufi scholars posited an 'inner' hermeneutics of the Quran called *ta'wil* (literally, 'bringing back to the root'), which allowed for multi-valent, non-literal, experientially-based interpretations of the Arabic text, adapted to express the needs of particular historical communities as well as the expressions of particular mystics.

In the historical development of Quranic exegesis, scholars distinguished between *ta'wil* and *tafsir*, the outer explanation of a passage. *Ta'wil* was primarily practiced in Shia and Sufi circles. Islamic scholar Annemarie Schimmel comments on the profundity of Quranic interpretation attempted by Islamic mystics and served by the Arabic language itself:

> [T]he mystics of Islam ... knew that a deeper meaning lies behind the words of the text and that one has to penetrate to the true core. It may be an exaggeration that an early mystic supposedly knew 7,000 interpretations for each verse of the Koran, but the search for the never-ending meanings of the Koran has continued through the ages. The Arabic language has been very helpful in this respect with its almost infinite possibilities of developing the roots of words and forming cross-relations between expressions.[8]

Like traditions of Jewish *midrash*, the Islamic traditions of mystical hermeneutics point to the importance of individual letters and letter-combinations. One tenet of early Ismaili *ta'wil* was that the written Quran was but a reflection of the 'Quran of creation', which itself contained the source of all symbols of the sacred. The Quran itself supports this interpretation by mentioning the 'Mother of the Book' (*ummil-kitabi*, Sura 43:4) and the 'Well-preserved Tablet' (*lauh mahfuz*, Sura 85:22), which remain with Allah in pre-existence.

---

[8]   Schimmel (1992), p. 48.

In the relation to this interpretation, modern Islamic scholar Seyyed Hossain Nasr relates the practice of *ta'wil* to Islam's unified cosmology of humanity, nature and the divine:

> In Islam the inseparable link between man and nature, and also between the sciences of nature and religion, is to be found in the Quran itself, the Divine Book which is the Logos or the Word of God ... It is both the recorded Quran (*al-Qur'an al-tadwini*) and the 'Quran of creation' (*al-Qur'an al takwini*) which contains the 'ideas' or archetypes of all things. That is why the term used to signify the verses of the Quran or *ayah* also means events occurring within the souls of men and phenomena in the world of nature.[9]

An important dimension of *ta'wil* is the confluence of spiritual experience and interpretation. As French Islamist Henry Corbin points out, the word *ta'wil* itself indicates 'an *exegesis* which is at the same time an *exodus*, a going out of the soul toward the Soul'.

> In Islam, *tawil*, 'the exegetic leading back to the source', answers to that law of interiorization, that experiential actualization of symbolic correspondences, which, being an innate and fundamental impulse of the religious Psyche, leads the Spirituals of all communities to the same goal.[10]

The same dense texture of sound and multi-valent letter roots, branching into multiple layers of meaning, also helps to support the notion the inimitability of the Quran (its *ijaz*), which according to Muslims is proof of its divine character as well as Muhammad's prophethood. On this basis, no literal translation into any other language is actually possible. As Schimmel points out, this led to the problem of how to transmit the contents of the Quran in lands where Arabic was not the native language, because 'not only the words and *ayat* but also the entire fabric of the Koran, the interweaving of words, sound and meaning, are part and parcel of the Koran'.[11]

Similarly, in prefacing his recent translations of the Quran, Michael Sells points out that considering the text in an oral, non-linear, communal context presents a key to its understanding:

> For Muslims, the Qur'an is first experienced in Arabic, even by those who are not native speakers of Arabic. In Qur'an schools, children memorize verses, then entire Suras. They begin with the Suras that are at the end of the Qur'an in its written form. These first revelations to Muhammad express vital existential themes in a language of great lyricism and beauty. As the students learn these Suras, they are not simply learning something by rote, but rather interiorizing

[9]   Nasr (1968), p. 95.
[10]   Corbin (1986), p. 134.
[11]   Schimmel (1994), p. 165.

the inner rhythms, sound patterns, and textual dynamics — taking it to heart in the deepest manner...

The Qur'anic experience is not the experience of reading a written text from beginning to end. Rather, the themes, stories, hymns and laws of the Qur'an are woven through the life of the individual, the key moments of the community, and the sensual world of the town and village.[12]

Using methods that emulate the context that creates *ta'wil*, a number of other recent authors have also attempted multiple, multi-valent translations of Quranic text into Western languages in order to communicate the dense intra-textuality of the original and to better communicate Quranic poetic language and concepts to non-scholarly Western audiences. These attempts have included 'open' translations, using multiple 'literal' and 'non-literal' poetic language.

For instance, the late Sufi author Lex Hixon created free renderings of important Quranic passages with the aim of communicating 'what, from my own experience in the world of Islam, the sensitive Muslim person actually feels when reading the Holy Koran or listening raptly, sometimes without clear verbal comprehension, to the melodious chanting of the classical Arabic'.[13] Like other Sufis or Ismailis who practiced *ta'wil*, Hixon rejected the distinction of 'literal' versus 'figurative' meanings of the text when considering the way in which the Quran is actually experienced by a Muslim, seeking rather to 'stay very close to the basic level of meaning in the Holy Koran'. He adds that 'What I would call the "basic meaning" of the verses is profoundly important. It forms the basis of Muslim practice and experiential belief, without which the various higher levels of mystical meaning would be nullified.'[14]

Some postmodern philosophies are in tension with hermeneutic approaches like *ta'wil* (or Jewish *midrash*). For instance, some[15] maintain that it is impossible to transcend our historical position in relation to a text and that there is no transcendental ego or awareness that can change this. From the standpoint of the mystical 'law of correspondences', which underlies much of Kabbalistic and Sufi thought, however, this simply represents the progressive de-sacralization of Western hermeneutics which could be seen to follow the same course as the de-sacralization of nature by Western science. If the world, or a text, has no inherent interiority, then a hermeneutics consistent with this view will consider it impossible to translate or communicate any meaning.

---

[12]   Sells (1999), pp. 11, 12.
[13]   Hixon (2003), p. 50.
[14]   *Ibid.*, pp 50–1.
[15]   Ormiston and Schrift (1990).

However, other postmodern voices may provide a bridge to *ta'wil* discourse. In particular, the so-called 'new paradigm' social science and educational research school has proposed theories that could be seen to parallel theoretical developments in the physical sciences, such as quantum mechanics and the principle of indeterminacy. For instance, Torbert, Reason and Rowan[16] propose research models that emphasize (1) a community process, (2) an open rather than closed field of research, (3) the development of an 'inter-penetrating' attention and (4) a spiral rather than a closed circle of hermeneutical inquiry. This school of research has questioned the view that interpretation, because it cannot be completely 'objective', must necessarily be completely 'subjective'.

Using the example of feminist approaches to history, they conclude that, while past experience cannot be transcended in making an interpretation, one can, by revealing this past experience as much as possible, open up an 'intersubjective' interpretation:

> Once this historicity of human experience is realized, it is clear that we must distinguish between some notion of an 'objective' understanding or interpretation which is unattainable and meaningless, and reach for an interpretation which is 'intersubjectively' valid for all the people who share the same world at a given time in history.[17]

Similarly, in proposing a model of collaborative research, William Torbert challenges the notion of 'controlled' research and criticizes much modern educational research as uneducational:

> Both in research and in organizational practice the effort at unilateral control presumes that the initial actor (whether researcher or practitioner) knows what is significant at the outset and that this knowledge is to be put to the service of controlling the situation outside the actor, in order to implement the pre-defined design as efficiently as possible.[18]

In such controlled research, if participants begin to question assumptions, examine methods or motivations, compare varying kinds of perceptual attention or otherwise depart from the researcher's plan, the research project is labeled 'out of control'. Torbert suggests that such 'controlled' educational research is not only 'anti-educational' in that it fails to discover anything new, but also anti-social in that it fails to prepare teachers or students for the world as it is.

As an alternative, Torbert suggests a model of 'action research' in which both the researcher and participants collaborate in an open system of 'experiments-in-practice' that are not rigidly controlled but, in

---

[16] See Rowan and Reason (1981) and Torbert (1981a, 1981b).
[17] Rowan and Reason (1981), p. 133.
[18] Torbert (1981a), p. 142.

fact, encourage the unexpected. In this respect, the setting of action research seeks to duplicate the conditions under which the research will eventually be applied — life itself.

In a number of elements, these new paradigm research models parallel the hermeneutics of *ta'wil*. The research of so-called unilateral control in the discourse of these researchers corresponds to the imposition of *a priori* religious (or academic) principles on the *ta'wil* interpretation or open rendering of a Quranic passage, including the experience of hearing it. In order to obtain one 'right' or 'objective' answer, suitable to all occasions, variables must be controlled and limited (for instance, some academic study requires a 'literal' rendering of a passage, but which one?). In both the new paradigm and *ta'wil* hermeneutics, the researcher or spiritual community becomes the central focus for inquiry and experience. The extent of control on the expression of *ta'wil* in a given community corresponds to the degree of control imposed upon the inquiry and on the range of what constitutes validity (that is, the 'usefulness' of the interpretation).

Besides encouraging an atmosphere of collaboration among participants, the action-researcher must, according to Torbert, develop an 'interpenetrating attention' capable of 'apprehending simultaneously its own dynamics and the ongoing theorizing, sensing and external eventualizing'. That is, this attention must bridge subjectivity and objectivity, neither discounting the researcher's own actions, feelings, thoughts and sensations nor allowing them to acquire so much importance that the rest of the system is lost to sight.

Torbert[19] suggests further that 'the prospective action scientist might well seek training in somatic movement forms such as tai chi, judo or the Gurdjieffian movements, all of which cultivate direct, moment-to-moment sensual awareness'.

Looking again at its parallels with *ta'wil* and *midrash*, Torbert's 'interpenetrating attention' corresponds to a contemplative awareness of word, meaning and symbol, influenced by both personal and community experience. Various Islamic and Jewish mystics over the ages developed practices that aimed to cultivate such an 'interpenetrating attention', using among other things meditation; concentration; breath and body awareness, visualization, and devotional chanting. They understood these practices within Islam and Judaism's shared doctrine of the unity of God, which mystics of both traditions interpreted radically as meaning that ultimately there is only one shared, sacred Reality.

---

[19]   Torbert (1981b), p. 443.

Finally, new paradigm research also proposes the concept of 'research cycles' proposed by Rowan.[20] In articulating the value of a 'dialectical paradigm' for research, Rowan recommends that new paradigm researchers begin to see their work as a spiral rather than a line or a closed circle. The stages of the research cycle or spiral, as Rowan outlines it, compare favorably to the process of multi-leveled translation of Biblical and Quranic texts pursued orally in mystical circles. *Being* corresponds to training in meditative awareness derived from the spiritual practices of the tradition. *Thinking* corresponds to the grammatical or language training necessary to approach a text. *Project* corresponds to the choice of text and the rendering of its Arabic roots in a thorough way. *Encounter* involves a confrontation with the text itself as a whole, by oneself and in relation to one's historical community of inquiry. *Making Sense* arises out of these multiple relationships. *Communication* involves translation of a text that opens meaning for another cycle of inquiry, beginning with the practice of *Being*.

The multiple renderings or interpretations of a particular sacred text spiral around its essential meaning, which can never be translated. However, the net of meaning that these multiple translations create places the reader or hearer within a symbolic universe that calls for his/her own experience to fix a final meaning for *this* moment, in relation to a particular community of inquiry. The indeterminacy of the interpretation can then be seen as a strength rather than a weakness, akin to the usefulness of the principle of indeterminacy in quantum physics.

The resulting meeting place between the language and philosophic concerns of Jewish-Islamic mystical ways of knowing and postmodern inquiry has suggested the notion of what I have called a 'hermeneutics of indeterminacy'.[21] Such a hermeneutic proves useful in considering particular Biblical and Quranic passages in translation and in opening up the conclusions that may be drawn from them. This hermeneutic has also proven useful in teaching Western students about comparative epistemology in the Bible and the Quran, comparing Jewish *midrash*, Christian theological exegesis and Islamic *tafsir* and *ta'wil*. Some of the features of the 'hermeneutics of indeterminacy' include:

1. *Structure and Openness.* There can be no one definitive translation or interpretation for all times, but several 'open', poetic translations can create an intersubjective bridge between the unique cultural, linguistic experience of a text and the experience of the interpretive community rooted in a different language.

---

[20]   Rowan (1981), p. 97ff.
[21]   See also Douglas-Klotz (1999).

2. *Multi-leveled, Evolutionary.* Each translation or interpretation can create a tapestry or net of possible meaning that can be meditated upon and interpreted according to the life experience of the person and community confronting it. The 'meaning', while rooted in the same text and participating in a reality connected to the historical religious experience of the community, reveals itself according to the needs of an emergent, evolutionary reality.

3. *Oral and Organismic.* In engaging in a community of inquiry, the written text leads the receptor toward the oral, both in its expanded translation style and in the encouragement to use methods such as guided listening, story-telling, chant and body prayer to experience the emotional and somatic dimensions of a particular text.

4. *Ecological and Relational.* The limitation of translation of sacred texts to one so-called literal translation can inhibit diversity and understanding in a learning context. In the context of a community that adheres to the notion of a unitive cosmology, the 'text' behind the sacred written text can be recognized as the manuscript of nature as it is experienced in a particular cultural, social, political and ecological con-text This is the system in which all study, interpretation and practice takes place.

20 years ago, I experienced all of these elements in a community of inquiry focused on the Quran by a septugenarian teacher from Pakistan. The following example of a 'hermeneutics of indeterminacy'[22] arose from my experience of a *ta'wil* of the Arabic of the first Sura of the Quran, Sura Fateha, and subsequent study of the rest of the Quran. All of the renderings below are 'literal' (based in the possible meanings of each word or root) but none become definitive until the reader adds her or his own relationship to the text to the process. Such a meditative rendering fulfills the root meaning of 'translation' — *carrying (meaning) across* — a linguistic and cultural bridge. At the same time, it tries to avoid becoming an object in itself in favour of evoking a response that engages the inquirer in a search for meaning. In the context of a learning community discussion, this search could be aided by contemplations and meditations that lead one back to two constants in the human experience: the awareness of the body and the awareness of nature. From a cultivated landscape of 'word-for-word' translation, the wilder aspects of the text's ecosystem then begin to re-appear.

The Opening

(Meditation on Sura Fateha, al-Quran 1 from the Arabic)

---

[22] Douglas-Klotz (1995), pp. 90–1.

Bismillahir rahmanir rahim

Upon hearing the Irresistible Voice of
Love's Wellspring and Goal,
we are led to affirm that

Alhamdulillahi rabbi-l'alamin arrahman irrahim

Whatever the Universe does, small or large,
through any being or communion of beings,
which helps further its purpose,
this act celebrates the Source of our unfolding story.
The essence of all praiseworthy qualities
constantly returns to the One Being
Give praise and celebrate!
This Being of beings mysteriously nurtures and sustains,
grows and brings to maturity
all worlds, universes and pluriverses,
all aspects of consciousness and knowledge,
all storylines and lesson plans.
This Source is the Original Womb of Love in all its aspects.

maliki yaumadin

It says 'I can' on the day when all elements part company and return
home,
when the threads of interweaving destiny unravel
and the invoices come due.
This Universe Being accepts the mission to resolve the unresolvable
at the time when time ends
just as it said 'yes' to the birth movements that began it.

Iyyaka n'abudu wa iyyaka nasta'ain

Cutting through all distractions, addictions and diversions,
all conflicting taboos, theologies, offenses and misunderstandings,
we will act only from this Universe Purpose,
we will develop abilities only in service to the Real,
we will bow to and venerate only the deepest Source of all Life
and we will only expect help from this direction,
the ration of what we need, freely given by the One.

Ihdina sirat almustaqim

We ask you to reveal our next harmonious step.
Show us the path that says, 'stand up, get going, do it!'
that resurrects us from the slumber of the drugged
and leads to the consummation of Heart's desire,
like all the stars and galaxies in tune, in time, straight on.

sirat alladhina an'amta 'alayhim

ghayril maghdubi 'alayhim wa laddalin.

The orbit of every being in the universe is filled with delight.
When each travels consciously,
a sigh of wonder arises at the expanse, the abundance.

This is not the path of frustration, anger or annoyance,
which only happens when we temporarily
lose the way and become drained,
roaming too far
from the Wellspring of Love.

Amin.

May this become the ground of our reality.

From Text to Life: The Sufi Traditions of State and Station

As implied above, a mystical way of knowing with regards to a sacred
story or text informs and is informed by a different way of considering
reality in general, both 'ordinary' and 'extra-ordinary'.

Today we find enormous popular interest in a 'mysticism of daily
life', as witnessed by the titles currently populating the self-help and
spirituality sections of most bookstores. The best of these books essentially tell us that whenever one is able to bring a certain quality of attention, concentration or love to a particular task, something like a mystical
insight can occur. Many others, however, concern attempts by various
authors to tell us that everyday life is really all the mysticism or spirituality or wisdom that there is. In other words, there is no purpose to
undergoing any sort of rigorous spiritual study or dedication to a practice, because the secrets of life are already here in plain view. They
essentially tells us not to worry about looking beneath the surface of
things and to simply float downstream with the tide of popular culture.
I would call this the 'Everything I Need to Know I Learned from Trying
to Figure Out How to Tune My VCR' thread in popular culture.

The earliest Sufi mystics did not attempt to justify reducing the rigour
of either prayer or spiritual practice. Instead, they take their lead from a
famous extra-canonical saying of Allah conveyed by the Prophet
Muhammad (one of the so-called *hadith qudsi*):

> My servant draws near to me through nothing I love more than the
> religious duty I require of him. And my servant continues to draw
> near to me by superogatory worship until I love him. When I love
> him, I become the ear by which he hears, the eye by which he sees,
> the hand by which he grasps, and the foot by which he walks. If he
> asks me for something, I give it to him; if he seeks protection, I pro-
> vide it to him.[23]

This saying, and other Quranic passages, provide what Carl Ernst has
called a 'divine charter for mystical experience' for the early Islamic and
Sufi mystics. Two examples of mystics who seem to have taken refuge
in the above hadith include Abu Yazid Bistami (d. 875, CE), who is

---

[23]   Translated in Ernst (1997), p. 51.

quoted to have said in a state of mystical ecstasy 'Glory be to me!' and Hussan ibn Mansur al Hallaj (d. 922, CE), who in similar state said 'I am the Truth' (*ana'l Haqq*).

These statements need to be understood in the light of a radical and uncompromising interpretation held by many early Sufis, of what is called in Islam the doctrine of Unity (*tawhid*). In this view, the expression of witnessing of faith, 'There is no god except God' (*la ilaha illa 'llah*) really means that there is no other reality except God. Hence any expression of 'I-ness' or individuality must be contained within this all-embracing Unity. Seyyed Hossain Nasr, expresses this idea in more nuanced language when he comments:

> Sufi doctrine does not assert that God is the world but that the world to the degree that it is real cannot be completely other than God; were it to be so it would become a totally independent reality, a deity on its own, and would destroy the absoluteness and the Oneness that belong to God alone.[24]

Early Sufi notions about mysticism and everyday life proceed from these metaphysical ideas about the complex interrelationships between and identities among human beings, the world and the divine, ideas that differ from classical Western notions, which usually dictate a division between the 'sacred' and the 'profane'. Many mystics accept that it is already the divine ground of Reality that has a mystical experience through them. Hence certain questions naturally arise: Why do these experiences pass away? How do they relate to the states of consciousness that remain? And what is the overall purpose of such experiences?

To try to answer these questions, some of the earliest Sufi authors begin to distinguish between *hal* (a mystical state) and *maqam* (a mystical station). The experience of a *hal* denotes a state of grace that descends upon a Sufi practitioner, but which is only temporary, facilitating (under ideal circumstances) a new 'station' in life, which represents the ability to bring a visionary state into everyday life.[25] For instance, Abd al Karim ibn Hawazin al Qushayri (d. 1074) comments:

> Among the folk, the state is a mode of consciousness that comes upon the heart without a person's intending it, attracting it, or trying to gain it — a feeling of delight or sorrow, constriction, longing, anxiety, terror, or want. States are bestowed; stations are attained. States come freely given while stations are gained with *majhud* (the expending of effort). The possessor of a station is secure in his station, while the possessor of a state can be taken up out of his state.[26]

[24]   Nasr (1991), p. 45.
[25]   Schimmel (1975), p. 99; Nasr (1991), pp. 72–3.
[26]   Translated in Sells (1996), p. 103.

Any tendency, however, to consider a *hal* or state as a mere phenomenon and therefore inferior to a station is countered almost immediatelly in Qushayri's discussion:

> If it did not change
> It would not be named a state
> Everything that changes, passes.
> Look at the shadow
> as it comes to its end,
> It moves toward its decline when it grows long.

> [S]ome of the folk have maintained the stability and perdurance of the states ... [T]hey are correct who claim that the state is continuous. The particular mode of consciousness (*ma`na*) is a taste or portion (*shirb*) in a person that can later grow into something more. But the possessor of such a continuous state has other states beyond those that have become a taste for him. These other states are ephemeral. When these ephemeral happenings become continuous for him like those previous states, then he rises up to another, higher and subtler state.[27]

In attempting to explain this qualitative difference in spiritual experiences or states, and the belief of early Sufis that not all states are equal, a well-known story is told about the first meeting of Shamsuddin Tabrizi and Jalaluddin Rumi (in approximately 1244 CE).[28] The story was written down about a century later by Shamsuddin Ahmed al Aflaki. In the story, Shamsuddin asked Rumi, 'Who was greater, Abu Yazid Bistami or Prophet Muhammad?' Rumi replies:

> 'Muhammad, God's envoy, is the greatest of mortals. What of Abu Yazid?'
> 'Then', said Shams, 'what does it mean that Muhammad said: "We have not known Thee as Thou shouldst be known," while Abu Yazid said; "I am exalted, my dignity is upraised, I am the sultan of sultans?"'
> Our Master replied: 'Abu Yazid's thirst had been quenched at one gulp; the jar of his understanding was filled with this little quantity; light was limited to the size of his window. But God's Elect sought each day further, and from hour to hour and day to day saw light and power and divine wisdom increase. This is why he said; "We have not known Thee as Thou shouldst be known."'[29]

Likewise, in attempting to differentiate the various stations, classical Sufi authors developed extensive lists of the various stages that aspirants would progressively experience as their everyday consciousness. One of the best known lists was formulated by the twelfth-century Persian Sufi Fariduddin Attar in his *The Conference of the Birds*. In this story

[27]   Translated in Sells (1996), p. 104.
[28]   According to traditions related by Bayat and Jamnia (1994).
[29]   Translated in Eflaki (1976), p. 21.

of a flock of birds who travel to meet their king (known as the 'Simorgh',), led by Solomon's sacred avian companion the hoopoe, the travellers must encounter seven valleys: the valley of the quest, the valley of love, the valley of understanding, the valley of independence and detachment, the valley of unity, the valley of astonishment and bewilderment, and the valley of deprivation and death. At the end of the journey, only thirty birds remain, and when they finally confront their 'king', they find their own image ('si-morgh' in Persian being a pun that also means 'thirty birds'.). The Simorgh tells them:

> I am a mirror set before your eyes,
> And all who come before my splendour see
> Themselves, their own unique reality:
> You came as thirty birds, not less nor more;
> If you had come as forty, fifty — here
> An answering forty, fifty would appear;
> Though you have struggled, wandered, travelled far,
> It is yourselves you see and what you are[30]

Other Sufi formulations add more complexity to the stations on the path, by again considering the question: who is doing the acting and experiencing? They do this by using the concepts of *fana*, an effacement or annihilation that involves giving up self-will, and *baqa*, the experience of realization-subsistence that involves finding the divine will in one's own.

For instance, the eleventh-century CE Sufi master Abu Said ibn Abi'l Khayr developed a list of progression of forty stations that a student must possess if 'his march upon the path of Sufism is to be acceptable'. This list begins with intention, conversion and repentence, includes effacement and subsistence as stations 21 and 22, then concludes with ascertaining of the Truth (38) , seeing God with the eye of the heart (39) and finally Sufism (*tasawwuf*) itself (as translated by Nasr 1991, pp. 78–82). Abi'l Khayr also relates each of the spiritual stations to a particular prophet, beginning with Adam and ending with Muhammad.

Viewed simplistically, this notion of a hierarchy of mystical stations accords with similar metaphysical notions previous and subsequent to the classical Sufis. To make matters more complex, however, as Nasr comments, Abi'l Khayr includes as stations what other Sufis called states, and he also includes additional stations after subsistence (*baqa*, the state of the 30 birds seeing themselves in the Simorgh). Nasr concludes his discussion on Abi'l Khayr's list on this point of seeming confusion, weaving the context of the state-station dialectic even more finely:

---

[30] Translated by Darbandi and Davis in Attar (1984), p. 219; see also the earlier translation of C.S. Nott, (1954) from a French version.

[T]he stations that follow [*baqa*, subsistence] may be said to be so many stations in the journey *in* God after the traveler has ended the journey *to* God. Even the station of service that comes after *baqa* must not be considered as action or religious service in the usual sense of the word but as service rendered by a being who has already tasted of union with God. In its own order it is something analagous to the vow of Avalokitesvara in Buddhism to save all creatures after having already set one foot in *nirvana*.[31]

The classical Sufis also began to develop the idea of the *waqt* or mystical moment, which Sells calls 'a time out-of-time within time, bringing the eschatological afterworld into the present'.[32] Qushayri comments in his treatise on mystical expressions:

They call the Sufi 'a son of his moment' meaning that he is completely occupied with the religious obligations of his present state ... He is concerned only with the present moment in which he finds himself. They also say: 'to be preoccupied with a past moment is to lose a second moment.'[33]

Based on this idea, Qushayri develops his ideas of deeper states. For instance, he sees what one might call the psychological or somatic experiences of the heart's constriction (*qabd*) and expansion (*bast*) as more developed than the beginner's concentration on an emotionalized hope of reward and fear of punishment, which are oriented to the future not the present:

The heart of the possessor of fear or hope is related to these two conditions through a deferring of the expected. But the possessor of constriction and expansion is a captive of his moment in the 'oncomings' that prevail upon him in the immediate now.[34]

It is as if the idea of the mystical moment as experienced by the experiencer creates a sense of time as pulse, including all times, rather than as objectified points on a line or circle. At this point in the conversation, the classical Western notions of the strict separation of past, present and future fall apart. We can see some of this as deriving from the predominately synchronic character of Semitic languages themselves and the difficulties even today of specifying past, present and future in them using only verbal forms.[35]

The doctrine of unity is also carried a step further by al-Hallaj in a text in which he even includes within the ground of the divine Reality what

---

[31]   Nasr (1991), p. 82.
[32]   Sells (1996), p. 100.
[33]   Translated in Sells (1996), p. 100.
[34]   Translated in Sells (1996), p. 106.
[35]   For more on the psychological dimension of the synchronic nature of Semitic languages, see Boman (1970) and a modern application to biblical studies in Douglas-Klotz (2000, 2003).

we might call negative states and stations, such as pride and disobedience. In retelling the story of Iblis' refusal to bow to the first human Adam (as told in Sura 2:30–3), Hallaj takes the side of the underdog and shows Iblis as the ultimate monotheist who will not bow to anything except the ultimate Reality, even when it is the ultimate Reality that orders him to do so. Iblis makes the argument in Hallaj's story that in order for the doctrine of *tawhid* (unity) to be true, Allah must have eternally foreknown and forewilled Iblis' disobedience and so he was, in a paradoxical sense, to use a colloquial expression, 'just following orders'. In a more profound sense, Hallaj sees Iblis as the perfect Lover who will not settle for anything except the divine Beloved, a theme explored by many later Sufis.

Through all of these formulations and paradoxes about *hal* and *maqam*, it is precisely a sense of divine Love that overshadows the need for any exact understanding, even while the attempt is being made. The Lover knows what s/he has experienced and does not need to articulate it in the words of the metaphysician. As Rumi says, commenting upon Quran Sura 62:5, intellect is like a donkey that carries books but does not understand them, whereas love is like the winged horse Buraq that carried Muhammad into the divine presence.[36]

Ordinary and Extraordinary: Extending the Conversation Across the Ages

We now turn to compare these insights with the writings of psychologists Reich and Maslow. Here we must acknowledge that there is both nuance and evolution in the thought of both on the issues explored above by the classical Sufis. To fully describe the thought of either writer would require separate studies devoted to each. At this point, I will content myself with some preliminary points of comparision, if only to show that deep thought on and experience of these issues sometimes produces similar results.

Maslow[37] puts forward both the concept of peak experiences (corresponding to the Sufi states) as well as what he called a 'hierarchy of needs' leading to full 'self-actualization' (the latter corresponding to the Sufi stations). For the early Maslow, an ecstatic type of unitive peak experience was essential for a fully self-actualized person, whom he defined as 'the fully growing and self-fulfilling human, the one in whom his potentialities are coming to full development, the one whose inner nature expresses itself freely, rather than being warped,

[36] Quoted in Schimmel (1975), p. 140.
[37] Maslow (1968).

repressed, or denied'.[38] Maslow called such an experience an example of 'B [for Being]-cognition': a way of knowing that was whole, complete, self-sufficient or unitary; that tends to de-differentiate figure and ground; that is open to a richness of detail as seen from many sides; that does not engage in comparison or competition; that seems outside of time and space, and in which the self is forgotten and the ego transcended.[39]

Later in life, Maslow re-evaluated his thought and began to consider the value of states that did not fit this early profile. In reconsidering his book *Religions, Values and Peak-Experiences*,[40] he wrote:

> [I] would now also add to the peak-experience material a greater consideration, not only of nadir experiences...but also of the plateau experience. This is the serene and calm, rather than the poignantly emotional, climactic, autonomic response to the miraculous, the awesome, the sacralized, the Unitive, the B-Values. So far as I can now tell, the high-plateau experience always has a poetic and cognitive element, which is not always true for peak experiences, which can be purely and exclusively emotional. It is far more voluntary than peak experiences are. One can learn to see in this Unitive way almost at will.[41]

At this point, he begins to approach the Sufi notions of an advanced station, that is, of a way of living ordinary life in an extraordinary way dependent upon regular practice and connection to a spiritual community:

> Very important today in a topical sense is the realization that plateau experiencing can be achieved, learned, earned by long, hard work. It can be meaningfully aspired to. But I don't know of any way of bypassing the necessary maturing, experiencing, living, learning. All of this takes time. ... The "spiritual disciplines", both the classical ones and the new ones that keep on being discovered these days, all take time, work, discipline, study, commitment.[42]

In the same re-evaluation, he also points out the dangers of polarizing the notion of 'mystic' versus 'normal', which some had done with his ideas. The danger of the 'polarizing mystic', he writes, is that such a person needs stronger and stronger stimuli ('triggers') in order to produce the same mystical state, which can lead to an addictive pattern. 'The great lesson from the true mystics, ... that the sacred is *in* the ordinary ... can easily be lost.'[43]

---

[38] *Ibid.*, p. 5.
[39] Maslow (1993), p. 89.
[40] Maslow (1964).
[41] Maslow (1993), pp. 335–6.
[42] *Ibid.*, pp. 336–7.
[43] Maslow (1993), p. 333, quoted in full above.

In this regard, Maslow's definition of 'transcendence' at the end of his life, itself transcended the traditional dichotomy of 'transcendent' and 'immanent' as usually discussed in the study of mysticism:

> Transcendence refers to the very highest and most inclusive or holistic levels of human consciousness, behaving and relating, as ends rather than means, to oneself, to significant others, to human beings in general, to other species, to nature, and to the cosmos. (Holism in the sense of hierarchical integration is assumed; so is cognitive and value isomorphism).[44]

Wilhelm Reich's way into the topic arises from his experiences treating both sexual dysfunction as well as schizophrenia using a type of breathing therapy. In his somatic theories, Reich describes a functional opposition of tension and charge leading to orgastic release. In many ways, this release corresponds to the states of the Sufis and Maslow's peak experiences. When the orgasm reflex was successfully reintegrated with the breath, Reich felt that there would be a reformulation (and/or release of) musular holding, which he called 'character armor'. This latter concept can be seen as the somatic equivalent of the Sufi station. For Reich, the analytic therapy of his mentor Freud was not fully effective, because it did not deal with this layer of holding:

> All of our patients report that they went through periods in childhood in which, by means of certain practices in vegetative behavior (holding the breath, tensing the abdominal muscular pressure, etc.), they learned to suppress their impulses of hate, anxiety, and love. Until now, analytic psychology has merely concerned itself with *what* the child suppresses and what the motives are which cause him to control his emotions. It did not inquire into the *way* in which children habitually fight against impulses ... It can be said *that every muscular rigidity contains the history and the meaning of its origin*.[45]

For Reich, such rigidity not only led to what he called 'character disorder', but also to a host of modern diseases such as cardiovascular hypertension, muscular rheumatism, pulmonary emphysema, bronchial asthma, peptic ulcer and various blood diseases. The restoration of a heathy orgastic response was itself a type of expanded state of awareness (a somatic *hal* in Sufi terms), which when consciously experienced led to a permanent change in the way a patient experienced the world (a somatic *maqam*):

> The sensation of integrity is connected with the sensation of having an immediate contact with the world. The unification of the orgasm reflex also restores the sensations of depth and seriousness. The patients remember the time in their early childhood when the unity

[44] *Ibid.*, p. 269.
[45] Reich (1948), p. 300, emphasis in the original.

of their body sensation was not disturbed. Seized with emotion, they tell of the time as children when they felt at one with nature, with everything that surrounded them, of the time they felt 'alive', and how finally all this had been shattered and crushed by their education.[46]

While considering in depth a case of the treatment of a schizophrenic patient, Reich makes a comparison between mystical states and schizophrenia, as viewed through the lens of his theory. This led him to propose that there was, in healthy individuals, the somatic equivalent of an organizing self, what he called an 'orgonotic sixth sense':

> Besides the abilities to see, hear, smell, taste, touch, there existed unmistakably in healthy individuals a sense of organ functions, an orgonotic sense, as it were, which was completely lacking or was disturbed in biopathies. The compulsion neurotic has lost this sixth sense completely. The schizophrenic has displaced this sense and has transformed it into certain patterns of his delusional system, such as 'forces', 'the devil', 'voices', 'electrical currents', 'worms in the brain or in the intestines', etc.[47]

In this regard, the difference between a mystical state and a pathological one, from Reich's viewpoint, depended upon a person's ability to integrate extraordinary states within her or his everyday life, to some degree or other:

> The functions which appear in the schizophrenic, if only one learns to read them accurately, are COSMIC FUNCTIONS, that is, functions of the cosmic orgone energy in undisguised form. Not a single sympton in schizophrenia makes sense if one does not realize that the sharp borderlines that separate *homo normalis* from the cosmic orgone ocean have broken down in the schizophrenic …
>
> I am referring here to functions which bind man and his cosmic origin into *one*. In schizophrenia, as well as in true religion and in true art and science, the awareness of these deep functions is great and overwhelming. The schizophrenic is distinguished from the great artist, scientist or founder of religions in that his organism is not equipped or is too split up to accept and to carry the experience of this identity of functions inside and outside the organism.[48]

Here Reich approaches the Sufi notion of *tawhid*, the doctrine of Unity, in which the knower experiences becoming the known. When the repression of sexual energy was magnified on the level of society, Reich saw what he called an 'emotional plague' that created war, dictatorship and all sorts of violence worldwide.[49] Later in life, Reich

---

[46]   Reich (1948), pp. 357–8.
[47]   Reich (1949), p. 454.
[48]   Reich (1949), p. 448, emphasis in original.
[49]   Reich (1949), p. 504ff.

believed that the ingrained repression of natural sexual energy was too deep to be changed in most people. The best that one could do was prevent the development of character armor in young children:

> If no severe damage has already been inflicted on it in the womb, the newborn infant brings with it all the richness of natural plasticity and development. This infant is not, as so many erroneously believe, an empty sack or a chemical machine into which everybody and anybody can pour his or her special ideas of what a human being ought to be. It brings with it an enormously productive and adaptive energy system which, out of its own resources, will make contact with its environment and *begin to shape that environment according to its needs...* LET THE CHILDREN THEMSELVES DECIDE THEIR OWN FUTURE. Our task is to protect their natural right to do so.[50]

For Reich like the Sufis, the natural expression of love seems to hold the key to doors that open a more integrative way of knowing as well as healing on both an individual and societal level.

As I interrupt this conversation between mystics across the ages, I find, in both the classical Sufis as well as Maslow and Reich, an attempt to contextualize 'everyday life' itself within a mystical framework, that is, not only is there a mysticism *of* everyday life, but everyday life is a type of mysticism in itself. Both endeavours, ancient and modern, attempt to bridge the inherent gap that ensues when a mystical or visionary state proves temporary, which could be expressed by the questions: If the way of knowing I'm currently experiencing is either divine grace or the essential nature of Reality, why does it go away? Or does it?

---

[50]   Reich (1983), p. 20, emphasis in original.

David Abram

Earth in Eclipse

There is another world, but it is in this one. (Paul Eluard)[1]

As a fresh millennium dawns around us, a new and vital skill is waiting to be born in the human organism. It is a talent necessary for any person who hopes to survive and flourish in the emerging era. We may call it the skill of navigating between worlds.

The twentieth century saw a remarkable multiplication of experiential domains, a rapid increase in the number of apparently autonomous realms with which we are forced to participate. There has been an astonishing proliferation of worlds, a diversification of separable realities, many of them mutually exclusive, that contemporary persons are increasingly compelled to frequent and familiarize themselves with. Each realm has its particular topology, its landmarks, its common denizens whom we come to know better the more we engage in that domain. Many of these realms hold a powerful attraction for those who visit them and, complicating matters, many of these experiential dimensions seem to claim for themselves a sort of hegemony, surreptitiously asserting their priority over all other dimensions.

Some of these worlds have been disclosed simply by the questing character of the human mind and imagination. Others have been made accessible to us through the probings of science, often in tandem with particular technological developments. Still others have been created for us by technologies operating more or less on their own. All of them now beckon to us.

Dissolving Distance

Among the most alluring of such worlds are those various experiential realms so recently opened by the electronic and digital media. Over a century ago, the invention of the telegraph and, soon after, the tele-

---

[1]   Quoted in Hirsch (1999), p. 9.

phone, made evident that geographical distance — the experience of near and far that determines our bodily experience of place — was no longer an absolute constraint. I, or at least my voice, could now make itself present somewhere else on the planet, as a friend across the continent could now make her thoughts heard over here, where I sit speaking to her. Our voices now found themselves wandering in a strange, auditory realm, often crackling with static, that existed beyond our ordinary, sensuous reality — a new experiential dimension wherein earthly distance and depth seemed to exert very little influence.

Another instance of this distance-less dimension was rapidly embodied in the radio, which brought into our home news of daily events happening far beyond the horizon of our everyday landscape — even as those very events were still unfolding! And then the television began to capture our gaze, its flickering glow in the den replacing the glowing flames around which families had traditionally gathered each evening. Through its blue screen, those events happening elsewhere became visible as well as audible realities. Other lands, other countries and cultures, suddenly became much more evident to us. We could no longer ignore them; they increasingly became a part of our lives, just as real as the spiders spinning their webs in the grasses outside, or perhaps — it seemed — even more real. But the television brought other worlds as well: the storied worlds of serials and soap operas whose characters became as familiar to many people as their own families. Or perhaps ... even more familiar.

In the final decade of the twentieth century another, more expansive, pasture opened up before us: the apparently fathomless labyrinth of cyberspace — a realm far more versatile and participatory than that inaugurated by either the radio or the television. Through the internet and the World Wide Web we seem to have dissolved terrestrial distance entirely, or rather, to have disclosed an alternative terrain wherein we can at last step free of our bodies and journey wherever we wish, as rapidly as we wish, to consult with other bodiless minds about whatever we wish. Or to wander, alone, in a quiet zone of virtual amusements. Instantaneous access to anywhere, real or imagined, is now available for collective engagement or for solitary retreat. Cyberspace, of course, is hardly a single space, but rather an ever-ramifying manifold of possible worlds to be explored, an expanding multiplicity of virtual realities.

The Allure of Transcendence

New as it seems, our fascination with the bodiless spaces made accessible to us by the digital revolution is only the latest example of our ever-expanding engagement with worlds hidden behind, beyond, or beneath the space in which we are corporeally immersed. One of the most ancient of such other worlds, and perhaps the first to exert a steady pull upon our attention, was the dimension of pure mathematical truths, the rarefied realm of numbers (both simple and complex) and the apparently unchanging relations between those numbers. Like great sea-going explorers setting out toward continents suspected but as yet unknown, mathematicians have continually discovered, explored, and charted various aspects of that kingdom whose lineaments, mysteriously, seem inexhaustible. The mathematical domain of number and proportion has long been assumed to be a separate, and purer, realm than this very changeable world in which we breathe and hunger and waste away — at least since the number wizard Pythagoras promulgated his mystical teachings in the city of Crotona some two and a half millennia ago — and the vast majority of contemporary mathematicians still adhere to this otherworldly assumption.

Pythagoras' faith that the realm of numbers was a higher world, untainted by the uncertainty and flux of mortal, earthly life, profoundly affected the thinking of the great Athenian philosopher, Plato, and through his writings this faith has influenced the whole trajectory of European civilization. In Plato's teachings, it was not just numbers and mathematical relations that had their source beyond the sensuous world, but also the essential character of such notions as truth, justice, and beauty; the ideal form of each such notion enjoyed the purity of an eternal and transcendent existence outside of all bodily apprehension.

Plato, that is, expanded Pythagoras' heaven of pure numbers and numerical proportions to include, as well, the pure and eternal "ideas" that lend their influence and guidance to human life. Indeed, according to several of the dialogues written by Plato for the students of his academy (in the fourth century BCE), every sensible thing — every entity that we experience with our senses — is but a secondary likeness of an immaterial archetype, or ideal, that alone truly exists. True and genuine existence belongs only to such ideal forms; the sensuous, earthly world, with its ceaseless changes, its shifting cycles of generation and decay, is but a corrupted facsimile of that truer, eternal dimension of pure, incorporeal ideals. That timeless domain cannot be perceived by the bodily senses: the reasoning intellect, alone, is able to apprehend that realm. According to Plato, the mind can never be fully at home in this bodily world; its true source, and home, is in that eternal realm of pure, disembodied ideas to which the rational mind secretly longs to return. Genuine reality, for Plato, is elsewhere.

As the intellectual culture of ancient Greece mingled with other cultures in the Mediterranean region, and as Pythagoras' and Plato's theories came in contact with the new religious impulses stirring on the edges of Hebraic culture, Plato's eternal realm of pure forms — ostensibly the true home of the intellect — provided the model for a new notion of eternity: the Christian Heaven, or afterlife. And as this new belief was given shape by the early Christian fathers, this eternity beyond the stars became the dwelling place not so much of the questing intellect as of the faithful and pious soul.

Today, the Heaven of Christian belief, together with the various heavens proper to other religious traditions, continues to exert a remarkable influence upon much of contemporary civilization. Even avowed atheists find their lives and their thoughts impacted by the collective belief in a heavenly realm presumed to exist radically outside of, or beyond, the palpable physicality of our apparent existence. Variously conceived as the afterlife, or as the dwelling place of God and his minions, such heavens are still assumed, by many, to be both the ultimate source of the sensuous world around us, and the ultimate end, or destiny, of our conscious awareness. Indeed, such transcendent realms still possess, for many of us, a clear primacy over the earthly world.

The Super-Small and the Ultra-Vast

But the ancient fascination with numbers was not only formative for the Christian notion of heaven; the mathematics it gave rise to also opened the way for the development of the secular sciences, and hence for the emergence of a host of abstract and increasingly otherworldly dimensions disclosed to us by those sciences.

One such realm powerfully impacting our lives today is the super-small dimension revealed by high energy physics: the subatomic world of protons and neutrons, of gluons and mesons and the elf-like quarks of which they are composed — and perhaps, underneath all these, the vibrating one-dimensional loops, or superstrings, that give rise to all such particles and their manifold interactions. Although very few of us have any clear apprehension of the subatomic world, or of the inscrutable particles that comprise it, we are continually assured by the physics community that this arcane realm is the ultimate source, or fundament, of all that we *do* apprehend. According to most contemporary physicists, the visible, tangible world glimpsed by our unaided senses is not at all fundamental, but is wholly structured by events unfolding at scales far beneath the threshold of our everyday awareness.

And yet physicists are not the only band of scientists inviting us to look askance at the world that we directly experience. According to a majority of researchers in the neurosciences, the perceptual world that envelops us — the world of rustling oak leaves, and bulldozers, and children racing through the spray of a loosened fire hydrant on a sweltering afternoon — is largely an illusion. Here, too, we must learn to recognize, underneath the apparent world, a more primary dimension: the realm of rapidly firing neurons, and neurotransmitters washing across synapses, of neural networks that overlap with other neural nets, an ever-shifting complex of interweaving neural patterns that continually generates — out of the endless array of photons cascading through our retinas and the sound waves splashing against our eardrums and the gradients of chemical molecules wafting past our nasal ganglia — the more-or-less coherent appearance of the surrounding world that we experience at any moment. Although we have absolutely no intuitive awareness of these events unfolding within the brain, our colleagues in the neurological sciences claim that such events provide the hidden infrastructure of all our perceptions. They insist that this unseen realm of synaptic interactions must be carefully studied and understood if we wish to truly understand just why the surrounding world appears to us as it does.

Meanwhile, in another set of laboratories, another group of intrepid researchers — molecular biologists tinkering with processes unfolding deep within the nuclei of our cells — have precipitated a growing cultural suspicion that the real and unifying truth of things, at least for organic entities like us, is to be found in the complexly coded structure of our chromosomes. Since the discovery of the double helix, molecular biology has become the dominant field within the life sciences, with a majority of its practitioners attempting to isolate the specific sequences of DNA that compose particular genes, and to discern the manner in which these genes are transcribed, by diverse chemical reactions, to generate the host of proteins that compose both the living tissues of any organism and the manifold enzymes that catalyze its metabolism. In giant, massively funded science initiatives, researchers race to fill gaps in the emerging map of the human genome, while others puzzle out the intricate epigenetic pathways whereby networks of genes interact to give rise to particular proclivities, dispositions, and behaviors. Numerous high-profit corporations devoted to the burgeoning technology of genetic manipulation are busy isolating or synthesizing the genes that ostensibly 'code' for desirable traits, eagerly transplanting them into various plants and animals in order to increase, presumably for human benefit, the productive yield of these organisms. Anyone even glancingly aware of these activities begins to suspect that the microscopic world of gene sequences and genetic interactions somehow determines our lives and our experiences. The ultimate source of our personality — of our habits, our appetites, our yearnings, and our decisions — would seem to be thoroughly hidden away from our ordinary awareness, carefully tucked within the nuclei of our cells.

We are often assured that such scientific worlds are entirely continuous with one another — that the subatomic world of protons and quarks is nested within the molecular world that makes up our DNA, that the DNA in turn codes, among other things, for the neuronal patterns that weave our experience. In truth, however, these worlds do not so easily cohere, for the arcane language that enfolds each of these dimensions is largely closed to the others. Many of those who speak the language of the brain sciences believe that their discipline holds the key to all that we experience, yet an analogous conviction may be found among many who speak the very different discourse of molecular biology and the genome, as do those other experts who traffic in the lingo of particle physics. It may be useful to assume that there are multiple keys to the hidden truth of the world, each key unlocking its own realm; yet the precise relation between these unseen realms — or the precise way to understand the relation between these realities — remains mysterious.

* * *

One of the most unnerving jolts to human experience occurred at the dawn of the modern era, when Nicolaus Copernicus offered a wealth of evidence for his theory — later verified by Galileo — that the fiery Sun, rather than the Earth, lay at the center of the visible universe. What a dizzying disclosure! The revelation that our Earth revolves around the Sun, rather than the other way around, ran entirely contrary to the evidence of the unaided senses, and it precipitated a profound schism between the sensing body and the reflective, thinking intellect. Suddenly, even the most obvious testimony of our senses, which daily reveal the slow course of the sun arcing across the sky and the stable earth beneath our feet, had been dramatically undermined. Henceforth a new, modern distrust of the senses, and of the apparent world revealed by the senses, began to spread throughout Europe. It is illuminating to realize that Descartes' audacious philosophical move, cleanly severing the thinking mind from the body — bifurcating reality into two, independent orders: thinking substance, or mind; and extended substance, or matter — was largely prompted by this new and very disturbing state of affairs. For in order to maintain the Copernican worldview, the thinking mind had to hold itself entirely aloof, and apart, from the sensing body.

Whether or not Descartes' ploy was ultimately justified, his conceptual unshackling of the cogitating mind from the body's world freed the modern intellect to explore not only the super-small realms of cells, atoms, and quarks, but also the ultra-vast spaces of star clusters and galaxies. In the last century we learned from our astronomers that the countless galaxies revealed by their powerful telescopes appear to be moving away from one another, and many of us have accepted, intellectually, the strange proposition that the universe is expanding. We've come to believe, quite matter-of-factly, in such logic-twisting phenomena as 'black holes', and in the rather confounding notion that, when looking up at a particular star in the night sky, we are in truth looking backward in time many thousands, or even millions, of years. Today, several of our most interesting and visionary astrophysicists and cosmologists suggest that this expanding universe is in truth only one of an uncountable plenitude of actually existing universes ...

An Outrageous Proliferation of Worlds

Thus, vying for our attention today are a host of divergent and weirdly discontinuous worlds. There is the almost impossibly small world of gluons and mesons and quarks, but also the infinitely vast cosmological field strewn with uncountable galaxies. We may be drawn to penetrate the electrochemical reality of neuronal interactions that presumably structures our psychological life, or perhaps to ponder and participate in the complexly coded universe of genetic transcription that lies at the root of all our proclivities and propensities, ostensibly determining the bulk of our behavior. Our desire may be stirred, today, not only by the religious heavens that many believe will supersede this world, or by the mathematical heaven of pure number and proportion toward which so many reasoning intellects still aspire, but also by the digital heaven of cyberspace — that steadily spreading labyrinth wherein we may daily divest ourselves of our bodies and their cumbersome constraints in order to dialog with other disembodied persons who've logged on in other places, or perhaps to try on other, virtual bodies in order to explore other, wholly virtual, spaces.

This proliferation of worlds — this multiplication of realities both religious and secular, super-small and ultra-vast, collective and solitary — is not likely to slow down in the coming decades. The accelerating pace of technological development seems to ensure that the proliferation of realms will continue to snowball. What *is* unclear, however, is whether the human mind can maintain its coherence while engaged in such a plural and discontinuous array, or disarray, of cognitive worlds. And if so, how?

Today, many persons rely on a kind of Alice in Wonderland strategy, taking one kind of cookie to shrink themselves down, another to make them grow larger — popping one kind of pill to deal with the mass of digital information they must navigate at their desk jobs, another to deal with their cranky children at home, and still another to withstand the daily onslaught of sadness and hype to which they subject themselves whenever they turn on the News:

> Astro-physicists are scrambling to account for the evidence, published this week in *Science*, that the universe is several billion years younger than had been assumed. Meanwhile, a contestant on *Survivor* has come down with a nose-bleed ...

> Government and industry leaders attending the landmark Earth Summit in Johannesburg today announced that efforts to slow global warming and other effects of pollution are woefully misguided: 'Of course technology is warming the planet,' said the CEO of Hubrizyne, 'but crucial advances in genetics now enable us to modify most species to fit the new planetary conditions!'

> Mothers! Are your children getting the most they can out of their sleep? Researchers have shown that children under the age of ten have more difficulty sleeping than they used to! Try SedaKind™, a newly patented sedative for children, now available in five flavors!

> ... An earthquake in the Caucasus has left several thousand people dead — but first, a message about your hair....

And so we tumble from one world to another, and from there to yet another, with no real translation between them: we slide straight from the horror of emaciated refugees running from the latest spate of ethnic cleansing to a bright and sparkly commercial for toothpaste. Turning off the television, we may practice tai chi for twenty minutes, tuning ourselves to the Tao, then step online in order to buy stocks in a gene-tech firm whose patented process for inducing Mad-Cow Disease in laboratory mice promises huge short-term returns.

The discontinuity — indeed, the sheer incommensurability — between many of the experiential worlds through which we careen on any given day, or which intersect the periphery of our awareness as we go about our business, entails a spreading fragmentation within our selves, like a crack steadily spreading through a china platter. We become increasingly multiple, without any clear way of translating between the divergent selves engendered by these different realms; we seek to draw our coherence from whatever world we happen to be engaged by at any moment. Or else we become numb, ensuring that no encounter moves us more than any other encounter, that no phenomenon impinges upon us more than any other, maintaining our coherence at the cost of our sensitivity and vitality.

How, then, can we find a way to move, to navigate between worlds without increasingly forfeiting our integrity, without consigning our minds and our lives ever more deeply to a kind of discombobulant confusion? School children fatally shoot their classmates without realizing there's anything wrong with an action they see played out a thousand times on the screen; parents, unnerved by such violence among the young, choose to raise their children in a strict fundamentalist manner, although the technological world that surrounds them confounds any literal belief in a transcendent creator. South of the equator, indigenous tribal communities lacking any notion of private property are abruptly plunged into the thick of modernity by the arrival of a television in their midst, or by the distribution of huge sums of money offered by corporations wanting to mine or develop their ancestral lands. ... Such stark instances, for which there exist no maps to help negotiate between discontinuous realities, mirror a disarray that is becoming ever more familiar to each of us.

When it does not immediately threaten our way of life, the proliferation of experiential worlds can also, of course, be deliciously exhilarating — a wild ride that regularly spurs us into an alert and improvisational responsiveness akin, perhaps, to that known by white-water kayakers, or by jazz musicians. Can it be, then, that we should accept and adapt ourselves to this ongoing state of dispossession and estrangement? Is it possible that such ceaseless realignment must now become our home — that it is time to welcome the steady slippage from one world into another, from one set of landmarks into another strangely different set, and from thence into yet another, exchanging horizons and atmospheres like we now change our clothes — becoming aficionados of the discontinuous and the fragmentary?

It is a tempting dream, but an impossible one. For in the complete absence of any compass — without a dependable way of balancing between realms — the exhilaration of steadily sailing from one wave-tossed medium into another cannot help but exhaust itself, giving way, in the end, to desperation, or to a numbed-out detachment void of all feeling.

Yet how, then, are we to find some equilibrium as we skid from realm to realm? How to orient ourselves within this deepening proliferation of cognitive worlds? Perhaps by paying attention to the patterns that play across these different zones, seeking subtle correlations, sniffing the air for familiar scents, striving to discern — hidden within this exploding matrix — the faint traces of a forgotten coherence. Perhaps by listening more closely we might glean certain clues to the way these diverse worlds conjoin. For indeed certain rhythms *do* seem to echo between these worlds, particular textures and tastes tug at the fringes of our awareness, reminding us of something ...

Only by such a process of attention can we start to discern the curious commonalities that are shared among these discontinuous dimensions. Only through such careful noticing are we brought to suspect that there is a specific space from whence all these common patterns were drawn — that among this wild profusion of worlds there is a particular world that has left its trace upon all the others. A unique province, indeed, that remains the secret source, and ground, of all these other realms.

A Common Source?

Yet how could this be? These worlds seem too incompossible, too incongruous, for them to be rooted in a common source. And how wierdly complex that source-world would have to be! If all these alien styles sprout out of the same land, how mysterious and downright *magic* would be that place!

And yet what a boon it would be to discover a specific scape that lies at the heart of all these others. For if there is such a secret world among all these — if there is a specific realm that provides the soil and support for all these others — then that primordial zone would somehow contain, hidden within its fertile topology, a gateway onto each of these other landscapes. And by making our home in that curious zone, we would have ready access to all these other realities — and could venture into them at will, exploring their lineaments and becoming acquainted with their denizens without, however, forfeiting all sense of orientation. We'd know that any world we explore remains rooted in that most mysterious realm where we daily reside, and so we could wander off into any of these other spaces without thereby losing our bearings; it would suffice simply to step back over a single threshold to find again our common ground.

But wouldn't it be common knowledge, by now, if there were a unique world that somehow opened directly onto all these others? Surely it would be a truth taught to us by our parents and professors as we gradually grew up into this dizzying situation! So we would expect … unless the one realm that impossibly joins all the others is the very realm that, traditionally, has been the most disparaged and despised; unless it is a zone that our religions all prefer to avoid, a dimension that our sciences all habitually overlook and forget. If this most remarkable world has conventionally been seen as the most derivative, drab, and problem-ridden of all the worlds, then perhaps our inability to notice this realm (and our reluctance to acknowledge its unruly magic) can be more readily understood.

Raindrops and Lettuce Leaves

The taken-for-granted world of which I speak is none other than the sensorial terrain of tastes and textures and ever-shifting shadows in which we find ourselves bodily immersed. Long derided by our religious traditions as a fallen and sinful dimension, continually marginalized by scientific discourse as a secondary, derivative, and hence ultimately inconsequential zone — how shall we characterize the sensuous world? It is the inexhaustible field of our most direct or unmediated experience, the very realm in which you now sit or recline, feeling the weight of your limbs as they settle down within the chair, or the stony bulk of the ground as it presses up against your flesh. It is the domain of smells wafting in from the kitchen, this place of rippling and raucous sounds, and of colors, too: the smudged white surface of the ceiling overhead, or the rumpled gray of the clouds gathering outside the window, their shadows gliding slowly across the road and the bending grasses.

This, in other words, is the body's world — that elemental terrain of contact wherein your tongue searches out a stray piece of lettuce stuck between your teeth, a fleshly zone animated by the thrumming ache within your skull and the claustrophobic feel of the shoes around your feet. Yet the sensuous world is animated by so much more than your own body; it is steadily fed by the body of the apple-tree, too, and by the old oak with its roots stretching deep into the soil, and the swollen bodies of the clouds overhead, and the warm, asphalt skin of the street, and the humming life of the refrigerator in the next room. This living, carnal field seems to breathe with your own moods, yet it's influenced, as well, by the rhythm of the rain now starting to pound upon the roof, and by the dark scent of newly drenched leaves and grass that drifts through the house when you swing open the front door. Shall we step out under those pelting drops to rescue the morning newspaper? Or perhaps that's too timid for a hot day — haven't we done enough reading? Why not toss this magazine into the corner, pry off our shoes and charge out under the trees to stomp and splash in the gathering puddles?

Why not, indeed? Lets do it! Since this — THIS! — is the very world we most need to remember: this undulating earth that we inhabit with our animal bodies! This place of thirst and of cool water, this realm where we nurse our most palpable wounds, where we wince at our mistakes, and wipe our tears, and sleepily make love in the old orchard while bird-pecked apples loll on the grass all around us — this world pulsing with our blood and the sap of pinon pines and junipers, awake with the staring eyes of owls and sleepy with the sighs of alley cats: this is the world in which we most deeply live.

Sadly, it is also the world we have most thoroughly forsaken.

Of course, we have not entirely lost touch with this place, where the moon slides in and out of the clouds, and the trees send down their thirsty roots, our nostrils flaring at the moistness of the night breeze. Our flesh calls us back to this earthly place whenever we are injured or sick, whenever we need to wash the dishes by hand, or clean out the overflowing roof gutters, or simply to empty our bladders and bowels on the toilet. Whenever we stumble and scrape our knees, when one of our tools breaks or one of our technologies breaks down, we must turn our attentions, if only for a moment, to the bothersome constraints of the gravity-laden earth that grips our bodies. Yet we'll linger only as long as we must; we know well that this messy world, with its stains and pockmarks and pimples, is not our destined kingdom. As soon as we've bandaged our knee, or repaired the dishwasher, or wiped our ass, we turn our attentions back toward those other, more compelling worlds. We turn back toward our laptop screen, or to the next page of the latest book on how to survive in the digital economy, or toward the churning sounds of a favorite audio disk pulsing out of our head-phones; we dial our colleague on the cellular to ask if she can join us at next week's conference on the newest gene-splicing techniques; or per-haps we plunge our attention back into our meditations on the tran-scendental unity hidden behind the apparent world. It never occurs to us that the most profound unity may reside in the very thick of the expe-rienced world itself, in the shifting web of interdependent relationships that ceaselessly draws the disparate presences of the sensuous cosmos, ourselves included, into subtle communion with one another.

The enveloping earth — this richly variant world alive with the sway-ing limbs of trees and the raucous honking of geese — is the very con-text in which the human body took its current form. Our senses, that is, have coevolved with the diverse textures, shapes, and sounds of the earthly sensuous: our eyes have evolved in subtle interaction with other eyes, and our ears are attuned by their very structure to the howling of wolves and the thrumming of crickets. Whether floating, for eons, as the single-celled entities that were our earliest biotic ancestors, or swimming in huge schools through the depths of the oceans, whether crawling upon the land as thick-skinned amphibians, or racing beneath the grasses as small mammals, or later swinging from branch to branch as long-tailed primates, our bodies have continually shaped themselves in delicate reciprocity with the manifold forms of the animate earth. Our nervous systems are thoroughly informed by the particular gravity of this sphere, by the way the sun's light filters down through the earth's sky, and by the cyclical tug of the earth's moon. In a very palpa-ble sense we are fashioned of this earth, our attentive bodies coevolved in rich and intimate rapport with the other bodies — animals, plants, mountains, rivers — that compose the shifting flesh of this breathing world.

Hence it is this animate, more-than-human terrain that has lent us our particular proclivities and gifts, our specific styles of behavior. The structure of our senses, our modes of perception, our unique habits of thought and contemplation, have all been formed by the mysterious character of the earthly cosmos in which we evolved. The sensuous earth thus provides the inescapable template for our experience of every other world we devise or discover. Whether we are plugging ourselves into cyberspace or simply synapsing ourselves to the page of a new novel, whether we are mathematically exploring the submicroscopic realm of vibrating, ten-dimensional strings or pondering the ultra-vast tissue of galaxies revealed by a new generation of radio-telescopes, we cannot help but interpret whatever we glimpse of these worlds *according to predilections derived from the one world in which we uninterruptedly live* — this bodily place, this palpable earth where we still breathe and burp and make love. Our intuitions regarding the lineaments of Heaven are inevitably shaped by those sensuous experiences that seem to correspond with such a place of equanimity and ease (luminous clouds drifting in the celestial blue, or a ray of sunlight that suddenly pours through a rent in those clouds and spills itself across a green hillside) and hence our religious heavens inevitably borrow their imagined structure from the evocative structures of this earthly cosmos. The way we envision the workings of DNA and the complex interactions between genes is similarly influenced by our encounter with the way things unfold at the scale of our most direct, unmediated experience of the sensuous earth around us. How could it be otherwise? It is our age-old encounter with the world at this scale that has provided the very eyes by which we peer through all our microscopes and telescopes, and that has fashioned the complex hands that now design our computer models.

Our comprehension of neural structures within the brain is deeply limited by the fact that the human nervous system did not evolve in order to analyze itself. The human brain evolved as a consequence of our bodily engagement with our earthly surroundings, and hence this brain has a natural proclivity to help us orient in relation to those dynamic surroundings. Whenever we attempt to focus the brain back upon itself, or upon any other hidden dimension — whether subatomic or galactic — it cannot help but bring those predispositions to bear, anticipating gravity, ground, and sky where they are not necessarily to be found, interpreting data according to the elemental constraints common to our two-legged species, yielding an image of things profoundly informed by our animal body and its accustomed habitat.

There is much to be gleaned from our investigations into other scales and dimensions, yet we consistently err by assuming our studies provide an objective assessment of the way those other realms really are in themselves. In order to convince ourselves of the rigor and rightness of our investigations, we consistently ignore, or overlook, the embodied nature of all our thoughts and our theories; we repress our carnal presence and proceed as though — in both our scientific and our spiritual endeavors — we were pure, disembodied minds, unconstrained by our animal form, and by our carnal and perceptual entwinement with the animate earth.

Thus do our sciences, like our religions, perpetuate the age-old disparagement of sensorial reality. The experienced earth lends something of its atmosphere to every world that we can conceive, and hence haunts these other worlds like a phantom. Each of the diverse and wierdly divergent worlds that cacaphonously claim our attention in this era — whether scientific or sacred, virtual or psychedelic, submicroscopic or meta-cosmological — is haunted by the animate earth. Of course most of our professors, priests, and scientists prefer not to confront such a vague presence that threatens, out on the very edge of our awareness, to disrupt all our certainties. Still, even the most confident scientists must sometimes find themselves wondering, late at night, how we can have gleaned such marvelous insights into the hidden structure of the universe while the most evident and apparent world that materially surrounds us seems to be choking and retching, its equilibrium dashed and its myriad plants and animals tumbling into extinction as a result of our human obliviousness. Can we really trust all our brilliant discoveries regarding the unseen causes that move the cosmos, when our own local cosmos is disintegrating all around us as a result of our collective inattention? Indeed, is it not likely that everything we think we know about *other* worlds has been distorted by our refusal to notice our thorough embedment in *this* world — by our refusal to discern the utter entanglement of our senses and our sentience within this breathing lattice of intertwined lives and living elements that we call earth?

Ethics and Otherness

Of course, the fragmentation and loss of coherence that we experience in our individual lives echoes a profound discombobulation within the larger community. In the absence of a common or broadly shared world, ethical instincts — including the mutual respect and restraint that hold a community together — steadily lose their grip, and indeed ethics, itself, comes to seem a largely arbitrary matter. When each of us expends so much energy and time engaged in worlds not shared by our neighbors (or even by other members of our own family), when we continually direct our individual attention to dimensions hidden above, behind, or beneath the shared world to which our senses give us access, it should come as little surprise that the *common* sense is impoverished, along with any clear instinct for the common good.

And yet an ethical compass — a feeling for what is right (or at least decent), and perhaps more important, for what is wrong — is especially crucial in such an era as this, when our technological engagement in other dimensions gives individuals a far greater power to manipulate experience, to violate others' lives and privacy, to inflict large-scale terror, and even to eradicate whole aspects of the real. Yet how is a genuinely ethical sensibility instilled and encouraged? How is an ethical sense (or, more simply, a good heart) born? Real ethics is not primarily a set of abstract principles; it is not, first, a set of rules (or "commandments") that can be memorized and then applied in appropriate situations. Ethics, first and foremost, is a feeling in the bones, a sense that there's something amiss when one sees a neighbor kicking his dog, that there's something wrong about hastening to one's work past a stranger who has tripped and fallen, her grocery bags torn, with their pears, cabbage heads, and a busted bottle of olive oil strewn along the sidewalk. Yet from whence comes the instinct to stop and help? The odd impulse to intervene with some teenagers stomping on an anthill, or simply to refrain from taking advantage of another's bad luck — where do these impulses come from?

It seems unlikely that the ethical impulse can be learned primarily from the pages of a book (not even from a text as deeply instructive as the Torah, or the Christian Bible, or the Koran). Still less can it be learned from the screen of a computer. For while these media readily engage the thinking mind, they cannot engage the whole of the thinking body (this reflective and sentient organism with its muscles and limbs) in the way that any face-to-face encounter, in the flesh, engages the whole body. It is in the flesh and blood world of our bodily actions and engagements that ethics has its real bearing. It is here, in this irreducibly ambiguous and uncertain land where we live with the whole of our beings — with our hands and our feet and our faces, with our bellies as well as our brains — that we are most vulnerable, most affected by the kindness of others, or by their neglect and disrespect. It is here, in this mortal world, that we are most susceptible to violence.

Of course we can strive to be basically responsible when engaged in those other, less palpable realms — for instance, when we are responding to a mountain of email, or cruising the internet, we can certainly try to respect electronic perspectives that are different from our own, or to refrain from violating the confidences and privacy of other participants. Similarly, we can attempt to be ethical in our experimental research with gene splicing and proteomics, or in our applications of nanotechnology, or in our electronic explorations of other planets. Yet unless we are already striving to act appropriately in our everyday, face-to-face interactions with the beings immediately around us, at the very scale in which we corporeally live — unless we are grappling with the difficult ambiguity of interacting with other bodily persons and presences without doing violence to those others — then we have no reason to trust that our more abstract, virtual engagements are genuinely ethical or good-hearted at all. For it is only in this palpable, earthly world that we are fully vulnerable to the consequences of our decisions and acts.

In those less sensorial dimensions (whether religious, scientific, or technologically mediated) we often find ourselves interacting with certain ideal presences that have been richly envisioned by our religious traditions, or with various provisional entities hypothesized by our fellow scientists, or with virtual beings invented and programmed for our entertainment. In such abstract, transcendent, or virtual worlds, in other words, we commonly encounter phenomena that may or may not be our own creation — we find ourselves interacting, there, with the manifold artifacts, interpretations, and projections of our own, richly imaginative species. But in the more immediate, palpable world to which our animal senses give us access, we encounter not only human creations but other *creative* entities — other persons as unfathomable as ourselves, and other earthborn entities whose sensations and experiences are even more unfathomable and mysterious. It is only here, in the earthly world of our carnal engagements, that we continually come into contact with beings (persons, deer, hawks) whom we can be certain are not primarily our own fabrications, but are really *other* — other centers of experience richly different from our own.

It is thus in the sensuous world that we most readily find ourselves confronted by what is genuinely, and indubitably, *other* than ourselves. It is only in this bodily terrain that we continually find ourselves in relation to other active agencies that are clearly not of our own making, to a world whose elemental lineaments we can be sure we did not devise. It is only here that we know we are in contact with what really exceeds us. And hence it is here, first and foremost, that ethical action really matters.

It is here, in the body's world, that an ethical sensibility is first engendered in any person. The seeds of compassion are sown in the palpable field of our childhood encounters with other sensitive and sentient bodies, in this ambiguous land where we gradually learn — through our pleasures and painful wounds, and through the howls and tears of others — to give space to those other bodies, discovering in their cries and expressive gestures a range of sensations strangely akin to our own, and so slowly coming to feel a kind of spontaneous, somatic empathy with other beings, and with our commonly inhabited world. It is this early, felt layer of solidarity with other bodies, and with the bodily earth, that provides fertile soil necessary for any more mature sense of ethics; it is this nonverbal, corporeal ability *to feel something of what others feel* that, given the right circumstances, can later grow and blossom into a compassionate life.

The child's somatic solidarity with others is inevitably a tentative and tenuous phenomenon, a layer of experience that emerges only when a child is free to engage, with the whole of her or his muscled and sensitive organism, in the animate world that immediately surrounds her. This quietly empathic layer of experience can arise, that is, only when the child is free to explore, at her own pace, this terrain of scents, shapes, and textures inhabited by other sensuous and sentient forms (by trees and insects and rain and houses), and so to discover, gradually, how to resonate with the other palpable presences that surround. It can arise only when the child is not deflected from such spontaneous, sensory explorations by being forced to engage, all too quickly, in the far more abstract and disembodied dimensions that beckon through the screen of a television or a computer.

How much violence has been done, in the latter half of the twentieth century, by planting our children in front of a screen! How many imaginations have been immobilized, how much sensorial curiosity and intercorporeal affinity has been stunted by our easy substitution of the television, with its eye-catching enticements, for the palpable presence of another person ready to accompany us on adventurous explorations of our mysterious locale?! The screen of the computer, too, requires us to immobilize our gaze, and to place our other senses, along with our muscles, out of play. It isolates and engages only a narrow slice of a child's sensorium, inviting her to set aside the full-bodied world that she shares with the fiery sun and the swooping birds. We should be profoundly skeptical about every exhortation by so-called experts to bring our children "on-line" as rapidly as possible in order to ensure their readiness and eventual competitiveness in the new "information economy." There is, of course, nothing wrong with the computer, nor with the many realms now so rapidly being opened for us by the astonishing capability of our computers — as long as we bring to these new realms the sensorial curiosity, creativity, restraint and ethical savvy that can only grow out of our full-bodied encounter with others in the thick of the earthly sensuous. But if we plug our kids into the computer as soon as they are able to walk, we short-circuit the very process by which they could acquire such creativity and such restraint.

Drinking the Rain

There can be little hope of renewing a collective sense of the ethical without beginning to acknowledge and honor the forgotten primacy of the one reality that we all have in common. Yet the only world that all humans have in common is the same that we share with the other animals and the plants: this terrestrial realm of wind and water and sky, shivering with seeds and warmed by the sun. Hence, it seems unlikely that we will locate a lasting ethic without rediscovering our solidarity with all these other shapes of sentience, without remembering ourselves to the swallows and the meandering rivers.

We are understandably fascinated by the rich promise of our technologies, and deliciously dazzled by the new experiential realms opened to us by the genius of the digital revolution. Yet our enthrallment with our own creations is steadily fragmenting our communities and our selves; our uncritical participation with technology risks eclipsing the sole realm that can provide the guidance for all our technological engagements. Indeed, only one realm is sufficiently wild, outrageous, and complex enough to teach us the use and misuse of our own creations.

Only by remembering ourselves to the sensuous earth, and rediscovering this place afresh, do we have a chance of integrating the multiple and divergent worlds that currently vie for our participation. Only by rooting ourselves here, and recovering our ageless solidarity with this breathing world — drinking the rain, feeling the fur on our flesh, and listening close to the wind as it whirls through the city streets — only thus do we have a chance of learning to balance, and navigate, among the proliferating worlds that now claim our attention at the outset of a new millennium.

There are many, many other worlds, yes, but they are all hidden within this one. And so to neglect this humble, imperfect, and infinitely mysterious world is to recklessly endanger all the others.

Anne Primavesi

Ecological Awareness

A Meeting of Science and Mysticism

In his *Moralia*, Plutarch tells the following story about the 'wise and good' wife of Pythes, a contemporary of Xerxes. It appears that Pythes came by chance upon some gold mines and was delighted, 'not with moderation, but insatiably and beyond measure', by their wealth. So he spent all his time at them, sent the citizens down into them and compelled all alike to dig or carry or wash out the gold, to the exclusion of all their other daily tasks. Many perished and all were exhausted.

The women then appealed to the wife of Pythes. She told them to go home and not lose heart. Then she summoned goldsmiths whom she trusted, secluded them and ordered them to make golden loaves of bread, all sorts of cake, fruits and whatever else in the way of dainties and food she knew Pythes liked best.

When he arrived home from abroad and called for dinner, she had a golden table set before him that held nothing edible. Everything was made of gold. At first he was delighted with the imitation food, but when he had gazed his fill, he called for something to eat. She served him with a golden replica of whatever he expressed a desire for. By this time he was angry and shouted that he was hungry. Then she said: 'But it is you who have created for us a plentiful supply of these things, and of nothing else. All skill in the trades has disappeared; the sowing and planting of crops has been forsaken and as no one tills the soil no food comes from it. Instead we dig and delve for useless things, wasting our own strength and that of our people.'

Not surprisingly, Plutarch recorded this tale under the heading 'Bravery of Women'. Pythes' wife shows award-winning courage in dealing with her autocratic husband. It may surprise us that she is not given a name: true to the social norms of his time Plutarch identifies her only as Pythes' wife. However this cultural gap between her and us shrinks when the story is read in the context of contemporary ecological awareness. Then her courageous actions become startlingly relevant to our own time as we appreciate not only the proclaimed moral of the tale but also its wider significance. And as we become conscious of the parallels between now and then, features of our own culture that require immediate attention are thrown into ever starker relief.

The first and most obvious parallel is that then and now, individuals' immoderate and apparently insatiable delight in monetary wealth can only be satisfied at the expense of others' welfare, human and non-human alike. As in the fourth century BCE, when individuals in our free market capitalist culture pay inordinate attention to the pursuit of wealth it means that many other groups are forced to work in intolerable conditions; or that they find themselves bankrupt, starving or out of work; and that as land and sea habitats are ravaged by industrial processes and machinery, our life support systems go into decline. Pythes is representative of a major corporate body today that neither recognizes nor acts upon moral criteria that would prevent it from harming others. Indeed it is legally compelled by pragmatic concern for its own interests to cause harm when the benefits to its shareholders of doing so outweigh the costs. The wider effects of this inordinate desire for wealth tend to be viewed as the inevitable (and therefore implicitly acceptable) consequence of corporate activity. In the technical jargon of economics they are 'externalities': that is, they affect a third party who has not consented to or played any role in carrying out a transaction. When an 'externality' affects people and the environment badly, executives have no authority to consider those effects unless they might negatively affect the corporation itself. In practice, that means ignoring any bad 'external' effects resulting from corporations' relentless and legally required pursuit of self-interest.[1] The effects are now manifest in the increase of many social and environmental ills worldwide, not least in the piles of inedible 'goods' that grow ever higher in refuse dumps sited far from corporate headquarters in the slums of South America, Africa, China and India.

Pythes' wife, once the women make her aware of the situation, deals with it effectively. This means that she finds a nonviolent way of causing her husband to *internalize* the effects of his actions. She makes him feel directly what it is to suffer hunger. In doing so she represents the ideal ecofeminist: someone who is aware of the connections between decision-making processes that authorize injustices based on gender,

[1]   Bakan (2004).

race and class and the exploitation and degradation of the environment. Ecofeminism's origins in antimilitarist direct action movements make its adherents attentive to the suffering inflicted on those who have to bear the social and environmental costs of the relentless pursuit of power and wealth. It also makes them alert to the need for and committed to the pursuit of nonviolent ways of redressing its effects.

Such a multi-issue, globally oriented awareness was sufficiently focussed in different groups by 1992, when the United Nations Conference on Environment and Development met in Rio, for it to impact on some of the documents that emerged from that meeting. They drew attention to four classes of people particularly affected by environmental degradation who are singularly powerless to do anything about it: women, children, indigenous peoples and the poor. Aware of this situation and of the fact that women belong to all four classes, groups of women and men worldwide have since developed and adopted different ways of remedying it. In general they are based on an acute sense of our common dependence on natural resources; of the importance of nonviolent relationships with each other and with nature and of the need for a broad gender base of knowledge, political power and skills within communities.

For Plutarch's story underlines what happens when any one form of awareness or attention is over-valued so that it dominates and therefore excludes others essential for preserving the richness of life. Pythes' attention to the wealth to be gained from gold dominated his life to such an extent that he lost all awareness of the consequences of exclusively pursuing that wealth. So he paid no attention to them. His wife's actions, however, show why she is presented as 'wise and good'. Her awareness of the beauty of gold and its ability to attract us is tempered by an awareness of the (literally) captivating effects of that beauty. The women whose lives were being ruined by it saw her as someone whose awareness of this, unlike her husband's, made her attentive to their needs and to those of their husbands, including her own.

Her wise response to the situation is, then, based on an integration of different levels of awareness tempered by an attentiveness to the effects of her actions on others as well as on herself. This ability to focus awareness on action beneficial to others flows from an attention towards their well-being that in this exemplary case is evoked by their distress. Based on an ever-expanding perception of the myriad of ways in which we relate to one another, it takes account of how this affects us at ever-deepening levels. In her case she related her husband's need for food to the needs of those made hungry by him. Then she used her perception of this connection between them in such a way as to raise his awareness of the fact that they were hungry, of the reasons for this and his role in it. Her action (one assumes) was beneficial to him and to them alike because underlying it was an even more profound awareness: that, deep in the earth and in the waters left unattended because of his actions, the roots of all living things embrace and as they do, support all life on earth including our own. Therefore she focuses his attention on *food* and on the fact that we are gifted with it by other living beings:

> From air and soil
> From bee and sun
> From others' toil
> My bread is won.[2]

This attentiveness lies at the heart of what I now call ecological awareness. Implicitly, it is a state of being: a quality that inheres in and belongs to a person alive to what gives all of us life and sensitive to the experience, at first or second hand, of what takes it away. Explicitly, it seeks ways to draw attention to the fact that it is the continuous interactions between earth's natural chemical, material and atmospheric cycles coupled with the relationships between them and living beings that sustain our life support systems — not gold, silver, dollars, oil or market shares. We emerged from those interactions, depend on them to sustain us during life and return to them in death. Built up over billions of years and only latterly affected by human activity, these resources cannot ultimately be quantified by us in monetary terms or properly absented from any moment of being alive. Nor can they be isolated in any real sense from what it is within the environment that relates us to each other in intersecting types of emotional, physical, religious, aesthetic and, increasingly now, technological response.

Given the gradual increase among us of this kind of explicit awareness, much of it stemming from the ecological sciences, we see its perennial truth reflected in Plutarch's tale. Pythes' attention to earth's resources is so taken up with an insatiable desire for what he sees as wealth as to effectively disengage him from every other aspect of those resources and from everyone and everything else. He has no emotional engagement with those around him that might lead him to attend to their welfare or remedy their suffering. His wife, however, once she becomes aware of the women's distress, engages positively with its causes and works to remedy the situation. She does so because she is aware that as her husband's action was intended by him to benefit no one but himself, by that very fact it has harmed many. And that as the act of feeding others is only made possible through the work, lives and deaths of many other beings, that is a moment to recognize their indispensable contribution to the good of many, husband and others alike.

---

[2] Primavesi (2000), pp. 160–1.

For that is how our interconnectedness and interdependence works. She is aware of this truth and attentive to its implications.

Her way of knowing what it means to belong within the wider community of life and to engage with it in a spirit of compassion and loving kindness is, I would hold, integral to any form of mysticism, whether theologically or scientifically conceived. I say this in spite of centuries of religious tradition in which individuals have been venerated as mystics on the grounds of practising what is usually presented as a world- and body-denying way of life. So in Plutarch's tale, Pythes would rightly be dismissed as 'unmystical' in his obsession with worldly wealth. But it is in its effects on him that we see just how unmystical he is. Mystics are properly identified by an awareness of, and a sense of communion with, that which transcends the individual and by doing so, binds us together into a greater whole. They understand themselves as part of that greater whole, however defined. Acting out of this often unarticulated sense of oneness, they know that what they are and what they do benefits (or not) both self and others. This sense of the whole, of the Holy, of all that there is, makes it impossible for them to consciously exclude any living entity from the remit of their concern. Instead, they are attuned to the well-being of all and to the effects of their actions on it.

So Andrew Harvey, in his collection of testimonies to the way in which mystics of different traditions have approached the 'unfathomable mystery' and sung its praises, urges us to recover 'the wisdom of the Mother', epitomized here in Pythes' wife. Harvey sees this wisdom as essential for 'a race in drastic denial of its interdependence': one that is therefore 'blindly devastating the environment'. What this wisdom brings to all of us, Harvey says, is

> the knowledge of inseparable connection with the entire creation and the wise, active love that is born from that knowledge. Unless the human race realizes with a passion and a reverence beyond thought or words its inter-being with nature, it will destroy in its greed the very environment it is sustained by.[3]

The Age of Ecology

While there have always been explicitly religious bases for this way of knowing, I want to look at those now offered us by the science of ecology. Through academic research and popular media presentations it has undoubtedly contributed to a worldview based on an implicit understanding of how we are inseparably connected with the entire creation. The terms 'ecology' and 'habitat', for example, are now household words used to discuss everything from ethics, shopping, politics

---

[3]　　Harvey (1998), pp. xii ff.

and agriculture to child-rearing. Such explicit scientific analysis of our interactive, interdependent global cohesion can be traced back to 1866 and the invention of both the term and the science of ecology by German biologist and philosopher Ernst Haeckel. The word 'ecology' combines the Greek *oïkos* (house, habitat) and *logos* (study, reason). The science of ecology means, Haeckel said, 'the entirety of the *science of the relationship of the organism with the environment*, including in a broad sense all the conditions of existence'. Later he offered a still broader definition in which he described the ecology of organisms as 'the knowledge of the sum of the relations of organisms to the surrounding outer world', including the correlations between all organisms living together in one locality.[4]

Ecology therefore presupposes that the whole of earth's living organisms constitutes a single, natural economic unit resembling a household or family. The economy in question is "Nature's economy", the natural one that exists to support the survival of individual organisms or species through a web of complex relationships with and within the global environment. There are no 'externalities' to it that can be discounted from the sum of life within it. No individual organism or species can survive independently of its relationships with others; and the interrelationships between their conditions of existence mean that life on earth for all organisms, including ourselves, is defined in terms of interdependence. Contemporary understanding of ecological interdependence is based on the principle that individual organisms exist only because of a shared dependence on others, even the most apparently insignificant. And that species' survival also depends on the place or niche each one occupies within the local environment and the contribution each makes to its welfare and to that of those around it.

The worldview briefly summarized here reflects an enlarged sense of awareness that is itself implicit in the language Haeckel uses. Ecology is '*the entirety*' of the science of relationships between organisms and the environment, of '*all*' the conditions of existence and the knowledge of '*the sum*' of relations between organisms and the surrounding outer world. As with Darwinism (Haeckel was an enthusiastic Darwinist), there is a definitive move toward including ourselves within that 'all', that 'sum', and he specifically urges researchers to keep the unity of the whole in mind.[5] Doing this consistently would, one assumes, raise awareness of ourselves as a dependent species within that whole: as one that belongs within the household of nature in the widest possible

[4]   Acot (1998), pp. 672, 703.
[5]   Acot (1998), p. 707.

sense and is therefore, like all those within it, dependent on the natural economy to sustain our existence, propagation and evolution.

Almost one hundred and fifty years after Haeckel, as this way of knowing ourselves mapped out by him becomes more and more scientifically explicit, our understanding of our proper place within that whole remains at best implicit. More generally, alas, its implications are ignored. Indeed, Andrew Harvey's uncompromising view of our situation finds such ignorance starkly displayed in its visible effects. All too clearly, for him as for many concerned scientists and citizens, our devastation of the natural environment demonstrates the drastic denial of our inter-being with nature.

This is the case in spite of the fact that the science of ecology has brought together, in an unprecedentedly explicit fashion, the knowledge of that environment and our place in it through interdisciplinary exchanges within a number of disciplines such as climatology, botany, geography, geology, animal chemistry and evolutionary biology. These exchanges have led to the development of internationally accepted and widely used concepts such as ecosystem, species extinction, climate change and biodiversity. Above all, the concept of 'environment' is used across disciplinary boundaries and invested with multiple meanings. So to discuss mysticism and ignore the environment in which its practitioners live and on which they depend is no longer a valid option for theologians or religious historians. By now routine use of ecological concepts should have created an explicit awareness that they too belong within and necessarily relate to the 'more-than-human' community of life, to use David Abram's pertinent phrase. And indeed in accounts of the lives of the Buddha, Jesus, Kabir and Aurobindo, the evidence as well as the expectation is there that their followers will relate to it in benevolent fashion.

But, while scientific data and ways of understanding them implicitly push us all now toward this ecological awareness, the pull of centuries of explicit philosophical and religious imaging and legitimizing of ourselves as being in some way distant from and distinct from the world, and as needing to distance ourselves from it in order to know God, has led to a stress on all that separates us (according to ourselves) from all other animals. Prime among these distancing devices has been our unique claim to being made in the image of God, and to that image being most clearly manifest in our faculty of reason. On that basis we have claimed to be created by God with the intelligence required to rule the earth. This notional distancing of ourselves from the earth has then served to justify objectifying it scientifically in order to use its resources solely for our benefit.

The potent mix of such self-perceptions, now largely implicit in secularized culture, remains all the stronger for being generally tacit. It is certainly strong enough for us to behave as if it outweighs the facts of our ecological interdependence. So we remain narrowly focussed on immediate human concerns in apparent isolation from those of the whole community of life. An important contributory factor to this mindset is the resistance among scientists themselves to any explicit scientific appeal to holism, such as that from Haeckel not to lose sight of nature as a whole: or that implied in James Lovelock's Gaia theory. To attend to such appeals would, environmental philosopher Karen Shrader-Frechette states, make ecology one of the most difficult and controversial sciences in which to achieve unifying and successful laws and predictions. This is so, she says, precisely because as we know more and more about communities of different species as well as living and non-living elements of the environment, we cannot point to any precise, empirically confirmed ecological whole. Also, because ecosystems for her are not agents in any meaningful sense, she thinks Lovelock, Rolston and others are wrong to suggest that they are holistic units engaged in maximizing their well-being. Her final objection to holistic explanations is that, despite their heuristic power, they are neither falsifiable nor even testable.[6]

These scientific reasons for rejecting scientific appeals to holism can be taken as typical of the difficulties such appeals pose for those within western culture who want to advocate a holistic worldview. For it is certainly true that we cannot experience, falsify or test any unit in nature as a whole: all our experiences of it and data culled from that experience are necessarily partial and fragmentary. But abstractions from those data to a framework within which they can be examined and discussed with others can, so to speak, help fill in the gaps between the parts under discussion. I take this relationship between an implied whole and observable phenomena or instances of behaviour to be one of ecology's strengths as a discipline.[7] Through such imaginative extensions of our observations of particular animals or plants and the relationships between them, and between them and us, ecological concepts such as that of an ecosystem can fill the gaps and make us aware of animals, plants and environmental conditions as an interactive, interdependent unitary whole.

So in his *Steps to an Ecology of Mind*, Gregory Bateson points to the cultural significance of such concepts. Lecturing students on diverse subjects, he found that the problem they had with learning from him was

[6]   Shrader-Frechette (2003), pp. 304, 308–10.
[7]   Primavesi (2003), pp. 70–86.

that they had been trained to think and argue *inductively* from data to hypothesis; whereas he tested hypotheses against knowledge derived by *deduction* from the fundamental concepts of science or philosophy.[8]. In a later book, *Mind and Nature: A Necessary Unity*, he argues for the relationship between data and hypothesis as that between what he calls the two great contraries of mental process, rigour and imagination. The hypothesis he invokes is that of connective pattern. 'What pattern', he asks, 'connects the crab to the lobster and the orchid to the primrose and all four of them to me?'[9] For him, the patterns can and must ultimately connect in the unity of the whole.

What characterizes this unity? Bateson's answer to this question takes us directly into the realm that would usually be termed 'mystical'. He uses the word 'sacred' to describe 'the whole within which we exist' and links that with another description: 'that with which thou shalt not tinker'. In other words, as sacredness is that which 'hooks up' or brings the whole together, any attempt to 'unhook' it or dissect it in order to manipulate it, control it or make it external to our existence is danger-ous for us: an admonition that takes account of the recorded occasions when 'the sacred' has rightly provoked not only awe and reverence but also fear.[10]

Throughout human history we have somehow been able to intuit that sacred unity, that wholeness. But precisely because of its wholeness, we have never been able to define it by anything other than itself. The whole is Holy: God is God; the absolute mystery is absolute. This is also the case because legitimate concerns about the partial and fragmentary nature of our experience rightly insist that we are limited in our ability to grasp 'the whole' in its totality. Keenly aware of these essential human limitations, philosopher Ludwig Wittgenstein gives us a neces-sarily paradoxical definition of 'the mystical' as: our 'contemplation of the world as a limited whole and the feeling of the world as a limited whole'.[11]

The most important aspect of the sacred then, and the one I want to stress above all, is that it cannot be separated out from whatever we understand as the whole of existence. In Bateson's words, we cannot 'tinker' with it; or if we try to, we do so at our peril. The sacred could, then, be described as the internal transcendence of every living being. But as that cannot be divorced from the environment and events in which each being is interconnected with all others, so we cannot reduce the sacred to any one manifestation of being but must extend the con-

[8]   Bateson (1972), pp. xvii ff.
[9]   Bateson (1985), pp. 16 ff.
[10]   Primavesi (2000), pp. 169 ff.
[11]   Wittgenstein (1961), p. 186.

cept to the whole dynamic system of relationships we call 'the world' and the relationship between that and God.[12]

Ecological Awareness and Theology

By using the word 'God' I have apparently moved on from a scientifically based mysticism to a theologically based one. I say apparently, because for me what Bateson names as sacredness and Wittgenstein as the contemplation and the feeling described as mystical cannot be divorced from the kind of ecological awareness detailed above. Nevertheless I recognize that even as ecological awareness slowly expands within western culture it is generally perceived as divorced from, or as making redundant our religious awareness of the unfathomable mystery theologians talk about as 'God'. The increase in one kind of awareness seems to signal, even to require a decrease in the other. But are they in fact linked in such a cause-effect relationship? Is it not rather the case that it is the explicit dimension of religious experience, when presented in the form of traditional 'God-talk', that has ceased to register as valid?

Danish theologian Lene Sjørup remarks rather ruefully that, not many years ago, it was almost impossible to say publicly in many parts of Europe that one did not believe in God. Today the situation is almost the reverse. To a large extent, the concept of the divine has been removed from the public sphere and relegated to private areas or to specific types of designated religious space, meetings, language and rituals. Yet on closer examination she finds evidence of a deep religiosity with strong social dimensions. Before going on to present her research into that religiosity and the ways in which people express it, she pinpoints one of the differences between them and conventional religious expressions that may have contributed to the 'Godlessness' of public life. She learned, she said, from theologian Sallie McFague that 'models of God are not definitions of God but likely accounts of experiences of relating to God'. And that all our 'religious' texts, including Scripture and the classics of the theological tradition, are records of interpreted experience.[13] For the most part, however, these records have been awarded the status of precise, authoritative definitions. So the precautionary theological principle that we can only record our experiences of God *as* a father, mother, beloved friend, host, spouse, love, truth or light is ignored by proclaiming that God *is* any or all of these.

Sjørup and many others she cites also make the obvious point that the 'recorders' in all the major world religions were almost exclusively men. Therefore most models of God in vogue today, whether or not

---

[12] Primavesi (2000), pp. 168 ff.
[13] Sjørup (1998), pp. 9, 23.

they are consciously taken as definitive, are based on male experience: on male ways of knowing and responding to God and to each other's claims to know God. In those formal bodies of religious faith growing today at a staggering rate, mostly outside northern Europe but affecting religious traditions there also, their adherents are all too often driven by a terrifying sense of being led by a God defined for them as Almighty, as Father and King; or as Judge condemning evildoers to eternal death. So they see themselves as engaged in divine combat against evil powers. In this combat, 'earthly' powers in every sense, and those who represent them, are to be sacrificed in God's name. And 'fallen Nature' is to be subdued, redeemed through plough and shears, pain and toil. Rachel Carson's carefully chosen epigraph for *Silent Spring* states pessimistically but realistically: 'Our approach to nature is to beat it into submission.' But does the word 'God' have to leave furrows of violence, traces of pain on the face of all living beings? And if so, how can it be used with religious integrity?

In today's increasingly literate society, such God-talk is, alas, made acceptable if not reinforced by Christian literalist reading of biblical texts. This means that they are read as if the (many times) translated words on the page can be attributed directly to God and indeed, as if God is bound by and within those printed words. So one bizarre American advertisement for the King James' translation of the Bible exhorts us: 'Read the Bible God reads'! All too often, what is then read into the biblical texts are interpretations of society and its environment that incorporates a perception of the natural world as profane and of certain man-made structures and buildings as 'holy'. This distinction presupposes, of course, that the sacred *can* be tinkered with by sequestering it within walls made by human hands.[14]

Because most western societies have evolved from patriarchal models, in nearly all of them their foundational religious texts contain no explicit reference to God as Mother. Or if they do, and it is brought to our attention, the weight of male reference within the texts and within their transmission so far outweighs the feminine as to make it literally unheard of. So in spite of efforts to heighten awareness of the female figure hymned as Wisdom in the biblical books *Proverbs* and *The Wisdom of Solomon*, the overwhelming focus on God as a male figure within all the other books and the God-talk based on that divine image remain deeply entrenched in western culture generally. Nobody knows the name of Pythes' wife.

Therefore Andrew Harvey's call for the recovery of knowing God as Mother as well as Father, and so recovering the wisdom of the Mother,

---

[14]   Primavesi (2004), pp. 55–58, 75 ff.

remains hard for us to hear and to respond to within mainstream religions. The excessively public nature of traditional religious institutions makes it difficult for those within them to share a personal, unmediated experience of the maternal divine with others, especially when its inwardness cannot be conveyed in conventional institutional terms. There is also the problem of how a specific contemporary, personal event can be assigned some more general truth or significance within a context that derives its own significance from past events and the authority of patriarchal voices.

Yet Harvey urges us to prioritize the recovery of that wisdom as a way of fostering a sense of inseparable connection with the entire creation and encouraging the wise active love that is born from the knowledge of that essential interconnectedness. This way of knowing is, I have tried to show, characteristic of real ecological consciousness. And whether or not it leads us to name our experience as mystical, it is one based on a sense of the sacredness of the whole and our oneness with it. One kind of awareness reinforces the other and so authenticates and increases our desire to act out of love for the benefit of the whole.

But how does this fit with conventional understandings of mysticism? Evelyn Underhill's *The Essentials of Mysticism*, a classic text on the subject, describes the central fact of the mystic's experience as 'an overwhelming consciousness of God and of his [sic] own soul: a consciousness which absorbs or eclipses all other centres of interest'. (To centuries of religious writers and readers, this reference to 'the soul', at least implicitly, denied 'the body' any role in this experience.) The mystic's communion with God, Underhill says, is always personal in the sense that it is communion with a 'living Reality, an object of love, capable of response, which demands and receives from him [sic] a total self-donation'. This sense of a double movement in which self-giving on the divine side answers to self-giving on the human side is, she says, found in all great mysticism. She sounds a cautious note against anyone speaking of mysticism who proposes as its object a reality that is less than God (such as increase of health or knowledge or happiness). Then, she says, we may begin to suspect that we are right off the track.

What does she see as the essentials of the mystic's way of knowing? For her the mystical way is best understood as a process of sublimation that 'carries the correspondences of the self with the Universe up to higher levels than those on which our normal consciousness works'. Normal consciousness sorts out some elements from others and constructs a certain order from them. But this order lacks any deep meaning or true cohesion. The claim of mystical consciousness, she says, is to a closer reading of truth: to an apprehension of the divine unifying principle behind appearance. She quotes Plotinus: 'The One is present everywhere and absent only from those unable to perceive it.' When we *do* perceive it, we 'have another life ... attaining the aim of our existence and our rest'. To know this, to be certain of it at first hand, is for her the highest achievement of human consciousness and the ultimate object of mysticism.[15]

It is clear that by using terms such as 'living Reality', 'the divine unifying principle' and 'the One', we have moved from restricting mysticism to the experience of communion between us and a 'male' God. It is also clear that such an understanding of mystical experience, while it doesn't exclude traditional models of God, cannot be restricted to them either: nor can our understanding of that experience be confined to traditional forms of theological discourse. This is not to say that they are missing from accounts of mystical experience. Lene Sjørup notes, for example, that among British people describing such experiences they often use the term 'the holy' for an image of the divine that inspires them with fear.

But a male God remains the most frequently used model, allied with a tendency to describe God, or the holy, as a supreme or higher form of intelligence: one that is omnipresent and omnipotent. This is understandable, since the most authoritative English biblical translation, the King James' version, belongs to an era when power was identified with being male and traditional metaphors of authority were taken from feudal societies. But, Sjørup says, these can hardly be upheld today in relatively democratic ones. So for her the old images of the divine lose credibility because of their dissonance with present images of authority. Though she herself doesn't make the connection between this and the reverse she noted in the numbers willing to acknowledge belief in God today, that connection surely exists. But what she does say is that in spite of the failure of old images to evoke the divine, the experience of the holy itself has not failed.

The exemplary models she cites — of oneness, boundlessness, wholeness, being enveloped in light or being part of a primal sea — not only embrace a wide spectrum of experience consonant with ecological consciousness. They also echo those examples offered by Evelyn Underhill, and again by Andrew Harvey, that belong within a wide historical and religious remit.[16] An important contextual change, however, appears in the historic gap between the experiences recorded by these last two authors. Whereas Underhill charts the record of mystical experience through early and late European literate culture and cites written accounts individuals have left us of their experiences of the divine,

---

[15]   Underhill (1920), pp. 2–7 ff.
[16]   Sjørup (1998), pp. 23 ff.

Harvey sets his account within a global context and begins by citing the voices 'of those original human cultures that lived in naked and reverent intimacy with nature'.

For most of us today, that experience of naked and reverent intimacy is not easily reclaimed, imagined or sustained. So the real question is whether or not a scientifically generated ecological consciousness can in some way help us reclaim or re-imagine it and then find a language to express it. In other words, are scientific concepts a way of knowing God that, by mediating a sense of reverent intimacy with nature and with its divine unifying principle, express the perennial mystical experience of communion between the self and God?

One way of answering this question would be to stress the fact that for me as a theologian, if God is to be God, then we cannot set any limits to the ways in which Godself is given in love to any being within the community of life on earth. Neither can we set any limits to or impose conditions on how each being might respond to that gift, nor restrict expressions of that response to certain terms, definitions or models. Rather, we need to find ways of describing that response that resonate with contemporary images of connective patterns between others and ourselves. It is also the case that as science discloses more and more of the wonders of life and the intricacies of interdependence within the whole of nature's economy, the limits of our knowledge and the limits of our language become ever more obvious. Therefore how could we confine the experience of communion with God within any particular set of human words?

I have seen this attributed to St. Francis and to the poet Rabindranath Tagore:

> I said to the almond tree: 'Speak to me of God!'
> And it blossomed.

Evident in these lines is the quality or virtue I call ecological humility: a quality that recognizes and reverences our kinship with every living being and each one's unique way of relating to and communing with God. Such an attitude, even when not overtly religiously based, is essential for those of us who seek to expand our ecological awareness with the help of science. The practical function of this humility is to remind us, as occasion requires, that whatever knowledge science offers us, it is always necessarily limited in its scope. Therefore whatever we claim to know about nature as a whole differs only in degree and in kind from what all its constituent beings know of it. And the same is true of our knowledge of God. We cannot claim for ourselves any additional faculty in regard to that way of knowing. We simply employ our faculties differently by focussing our attention on different aspects of our experience.

The scientific question then is how we know anything about nature. We know about it, according to many eminent philosophers and scientists, because we are rationally privileged, uniquely capable of 'intelligent information processing'. This argument for human supremacy based on our 'rational privileges' usually assumes that, given time, we are capable of intelligent observation of the 'entirety' of nature and therefore could draw the 'right' conclusions from those observations.[17] But as Shrader-Frechette reminds us, here too our inadequacy is all too evident. Especially when confronted with proof of that inadequacy in the visible and devastating effects of today's military-industrial privileging of scientific knowledge. And are we generally aware that, more than two thousand years after Pythes, gold mining today generates twelve tons of waste when extracting enough gold to make a ring?

Yet awareness of the need for a precautionary attitude to our ways of knowing ought not to prevent us acknowledging our particular human gifts in that respect. Rather we should use them to develop a sense of the world as gift: and rejoice in our ability to make explicit a sense of loving communion with its Giver:

> My songs are the same as are the spring flowers,
> they come from you.
> Yet I bring you these as my own.[18]

Tagore's lines remind us that whatever kind of knowledge of God we bring to our relationship with God is, in the first instance, God's gift to us. And furthermore, that each living entity responds to that in a manner unique to itself. Today science offers us the religious opportunity to make explicit a sense of the world as sacred; by combining new data with hypotheses and by using imagination with rigour. But always with the proviso, as Evelyn Underhill reminds us, that we do so in the hope of entering into communion with God. Our attention then is not focussed, as Shrader-Frechette declared, on achieving 'unifying and successful laws and predictions' that would merit a Nobel Prize in science. For knowing God is not an achievement on our part but a divine grace and favour: it is a lifework in which our ignorance will always outweigh our knowledge.

This for me is the point at which we come to understand that our ways of knowing God and the world must translate into something other than inadequate concepts. So Thich Nhat Hanh urges us 'to develop understanding in order to be able to love and to live in har-

---

[17] Primavesi (2003), pp. 30–8.
[18] Tagore (1921), p. 106.

mony with people, animals and plants'.[19] In similar fashion the poetic Marxist axiom declares that 'an inner light' kindles kindness that spreads outward like a consuming fire.

In Pythes' wife we saw an exemplary form of such kindness: one that made her attentive to the suffering of others and able to act in such a way as to benefit them. This outward attention stemming from an enlightened awareness is a constant element in all religions. It is evident in the life and teachings of the Hebrew prophets, of Jesus and of the Buddha, each of whom asks for exemplary care and attentiveness to the suffering of those classic characters: the poor, the sick, the imprisoned, the naked, the hungry and the stranger in the land. Now the expansion of ecological consciousness teaches us to extend that awareness of suffering beyond our species to all those affected by our actions. And then, as Thich Nhat Hanh exhorts us, not only to be aware of the suffering but, out of that awareness, to cultivate love and compassion and learn ways to protect the lives of people, animals and plants.

This echoes the words of the prophet Micah:

> [God] has shown you what is good: what is asked of you.
> Act justly; cherish compassion and loving kindness;
> Walk humbly with your God.[20]

Such mystical voices speak to us today of a reverent intimacy with nature, one born out of a deepening understanding of our essential 'inter-being' with all its inhabitants. They tell us of the necessity of profound respect for everything that lives; of a peace that is the birthright and the task of all who honour the Great Web of Life; and of the urgency of humility before the majesty of the universe.[21]

---

[19] Nhat Hanh (1987), p. 64.
[20] Micah 6.8.
[21] Harvey (1998), pp. 1–2.

Chris Clarke

Final Reflections

The vital practicality of all these ideas emerges time and again from most of the chapters. Spirituality, or rather its perversion, has become a pivotal element in today's world, where the dogmatic, violent and intolerant factions of religions are on the increase — from fundamentalist Christians in the USA, to Al Qaida and Shas in the Middle East, to the BJP in India. The particular routes through which this distorted spirituality arises stem from complex political factors — in some places it is a grass-roots rebellion of the poor, but more often (as in the USA) it is a manipulation of feelings of fear by right wing political factions. But the ultimate source is the same: it is the shadow side of the triumphant rationality of the West, a rationality which has cut itself off from mystical knowing, and repressed it along with the economic repression of the poor. Reliance on the simplistic black and white logic of the propositional/analytical side of our mind then polarises views across the world, whether it is between rich and poor or conservative and liberal, into seeing the world as divided into the camps of good and evil, leading to a vicious cycle of increasing fear. The relational/mystical side of our mind, cut off from spiritual understanding, responds to fear and the threat to the self by generating an overpoweringly passionate certainty in each group's own religious experience, coupled with an equally passionate damning of those with alternative experience. The two central meaning-making systems of the mind are out of joint, resulting in escalating cycles of aggression and violence.

The way out of this cycle involves the re-integration of our thinking, honouring diverse ways of knowing while being open to the constant growth and change that flows from the Spirit. It involves integration of the ways of knowing that have become sundered, and integration of the parts of society that carry the riches of each of these ways. It has to recognise those whose faithfulness to other ways has led to their repression by the dominant society — the women and the indigenous peoples, the *anawim* , the downtrodden, to use a term from Jewish and Christian spirituality[1] — and the unforgiven in our asylums. We have known this for many years, through the analysis of the nature of inter-community dialogue by the physicist David Bohm, through the work of pluralist philosophers of science such as Nancy Cartwright, through the work of feminist theologians such as Mary Grey. Two things are new: first, an understanding of precisely how our failure or our success in achieving this integration is embodied in the structure of our minds; and second, an understanding that we ourselves are cocreators of our world through our spirituality, or through its distortion.

Faced with the entrenchment of our divisions, is there any real hope of achieving the transformation that is needed? It will certainly not come without the courageous engagement of those open to spirit. But now the twenty-first century is starting to reveal the first signs that offer a possibility of moving towards what Mike King has called the 'post-secular society'.[2] Let me end by speculatively listing some of these signposts to our healing.

- Spirituality is gradually starting to be accepted into mainstream education. In the UK it is now part of the national curriculum[3] and in starting to appear in higher education mission statements.[4]

- Also in the UK, spiritual considerations have started to enter the UK mental health services: for example, the growth of the 'recovery model'[5] is making awareness of spiritual issues a part of the training of some therapists.

- There is growing awareness of the large proportion of people within Western societies who are open to spiritually based values but find no means of expression through a spiritual community — the 'cultural creatives' identified by Paul Ray some ten years ago.[6]

- There is a preliminary rapprochement between spirituality and science, vital because in our society science has secured the high ground of political influence. While there is still much pseudo-science that can only alienate the scientific community,

---

[1] E.g. Psalm 147.6 'Yahweh sustains the humble [*anawim*]but casts the wicked to the ground.'
[2] King (2002, 2003).
[3] Best (1996), p. 35.
[4] E.g. University College Winchester, www.winchester.ac.uk.
[5] Ralph. and Kidder (2000).
[6] Ray and Anderson (2000).

progress is represented both by high quality popular books[7] and prestigious academic books.[8]

While these signs suggest the first openings of a route to a post-secular society marked by tolerance and spiritual values, including justice for the planet and all beings that live upon it, and while I believe that such a transition is a practical possibility, I do not think that it will happen either smoothly or automatically. The route to the post-secular society lies through overcoming the conflict and fear of our divisions by non-violent means. The process is well documented. Many spiritual paths teach that it is love that casts out fear within the individual; and within society the non-violent means of overcoming conflict named by Gandhi as *satyagraha,* (literally: holding firmly to truth) considered by him as 'a relentless search for truth and a determination to reach truth … a force that works silently and apparently slowly. In reality, there is no force in the world that is so direct or so swift in working.' (From a letter, 25.1.1920)

---

[7]   E.g. Ravindra (2000).
[8]   E.g. Russel et al. (2001).

References

A Course in Miracles (1975). Glen Ellen, CA: Foundation for Inner Peace,.

Acot, P. (ed.) (1998). *The European Origins of Scientific Ecology (1800-1901).* Amsterdam, Overseas Publishers Association.

Aldridge, David (1996). *Music Therapy Research and Practice in Medicine — From Out of the Silence.* London: Jessica Kingsley.

Andrews, Lyn D. (1999). *The Cosmic Connection and the Theory of Universal Fulfillment,* www.philosophicfriend.org.

Assagioli, Robert (1994, first published 1974). *The Act of Will.* London: Aquarian/Thorsons.

Attar, Farid ud-Din (1971). *The Conference of the Birds (Mantiq ut-Tair).* Trans. G. de Tassy and C.S. Nott. Boulder, CO: Shambhala.

Attar, Farid ud-Din (1984). *The Conference of the Birds,* trans A. Darbandi and D. Davis. London: Penguin.

Bache, Chris (2000). *Dark Night, Early Dawn: Steps to a Deep Ecology of Mind.* Albany: State University of New York.

Bakan, J. (2004). *The Corporation,* London: Constable.

Ball S. J. (ed.) (1990), *Foucault and Education: Disciplines and Knowledge.* London: Routledge.

Barnard, Philip (2003). 'Asynchrony, implicational meaning and the experience of self in schizophrenia', in T. Kircher & A. David (eds.) *The Self in Neuroscience and Psychiatry* (pp. 121-6). Cambridge: Cambridge University Press.

Bateson, Gregory (1972). *Steps to an Ecology of Mind.* New York, Ballantine Books.

Bateson, Gregory (1985). *Mind and Nature.* London, Flamingo.

Bayat, Mojdeh and Mohammad Ali Jamnia (1994). Boulder, CO: *Tales from the Land of the Sufis.* Boston: Shambhala.

Belenky, Mary Field; McVicker Clinchy, Blythe; Rule Goldberger, Nancy and Mattuck Tarule, Jill (1986). *Women's ways of Knowing: The Development of Self, Voice, and Mind.* New York: Basic Books.

Beena, C. (1990). *Personality Typologies: A Comparison of Western and Ancient Indian Approaches.* New Delhi: Commonwealth Publishers.

Berman, Morris (1994), *The Reenchantment of the World.* Ithaca, London: Cornell University Press.

Best, Ron (ed.) (1996). *Education, Spirituality and the Whole Child.* London: Cassel.

Bickham, Jack (1992). *The Writer's Digest Handbook of Novel Writing*, Cincinnati: Writer's Digest Books.

Boman, Thorlief (1970). *Hebrew Thought Compared with Greek*. New York: W.W. Norton.

Bomford, Rodney (1998). 'Mapping Mental Processes', in *Journal of Melanie Klein and Object Relations*, **16**, 1, March 1998, pp. 35–46.

Bomford, Rodney (1999). *The Symmetry of God*. London: Free Association Books.

Boyce-Tillman, June B. (1996). *In Praise of All-encircling Love II*, London: Hildegard Press and The Association for Inclusive Language.

Boyce-Tillman, June B. (2000). *Constructing Musical Healing: The Wounds that Sing*. London: Jessica Kingsley.

(*see also* Tillman)

Brown, D.P. (1986). 'The Stages of Meditation in Cross-Cultural Perspective', in *Transformations of Consciousness: Conventional and Contemplative Perspectives on Development*, eds. K. Wilber, J. Engler, and D. Brown, 219–84. Boston: Shambhala.

Bruteau, Beatrice (1997). *God's Ecstasy: The Creation of a Self-creating world*. New York: Crossroad publishing.

Bucky, Peter A. and Weakland, Allen G. (1992). *The Private Albert Einstein*, Kansas City: Andrews and McMeel.

Buswell, R.E. and R.M. Gimello (1992). 'Introduction' in *Paths to Liberation: The Marga and Its Transformations in Buddhist Thought*, eds. R.E. Buswell and R.M. Gimello, 1–36. Honolulu: University of Hawaii Press.

Campbell, Joseph, (1960–8). *The Masks of God: [Vol.1] Primitive Mythology (1960); [Vol.2] Oriental Mythology (1965); [Vol.3] Occidental Mythology (1965); [Vol.4] Creative Mythology 1968*. London: Secker & Warburg

Campbell, Joseph (1972). *The Hero with a Thousand Faces*. New Jersey: Princeton University Press.

Christian, W.A. (1972). *Oppositions of Religious Doctrines: A Study in the Logic of Dialogue Among Religions*. New York: Herder and Herder.

Clarke, Chris (1996). *Reality Through the Looking Glass*. Edinburgh: Floris Books

Clarke, Chris (2001). 'Construction and reality: reflections on philosophy and spiritual/psychotic experience', in Clarke, I (2001) pp. 143–62. See below.

Clarke, Chris (2002). *Living in Connection: Theory and Practice of the New World-view*. Warminster: Creation Spirituality Books.

Clarke, Isabel (ed.) (2001). *Psychosis and Spirituality: Exploring the New Frontier*. London: Whurr Publishers Ltd.

Clarke, Isabel, (2001a). 'Psychosis and Spirituality; the discontinuity model', in Clarke, Isabel (ed.) (2001) See above.

Cohen, Leonard (1992). 'Anthem' from the CD *The Future*. Columbia

Coomaraswamy, Ananda K. (1985). *The Dance of Siva*, New York: Dover Publications.

Corbin, Henry (1986). *Temple and Contemplation*. London: Islamic Publications.

Cowling, Jamie, (ed.) (2004). *For Art's Sake. Society and the arts in the 21st Century*. London: Institute for Public Policy Research..

Dalai Lama, H.H. (1988). *The Bodhgaya Interviews,* ed. J.I. Cabezón. Ithaca, NY: Snow Lion.

Dalai Lama, H.H. (1996). *The Good Heart: A Buddhist Perspective on the Teachings of Jesus.* Boston: Wisdom Publications.

Danchin, Lauren (1989). *Raw Vision.* Issue 1 Spring 1989

Dean, T. (ed.) (1995). *Religious Pluralism and Truth: Essays on Cross-Cultural Philosophy of Religion.* Albany, NY: SUNY Press

Hayward Gallery (1979). *Outsiders. An art without precedent or tradition.* London: Arts Council of Great Britain.

Dewey, John (1910). *How We Think.* New York: Heath and Co.

Douglas-Klotz, Neil (1984). 'Sufi Approaches to Transformational Movement', *Somatics,* 5, No. 1 p. 44

Douglas-Klotz, Neil (1995). *Desert Wisdom: The Middle Eastern Tradition from the Goddess through the Sufis.* San Francisco: HarperSanFrancisco.

Douglas-Klotz, Neil (1999). 'Midrash and postmodern inquiry: suggestions toward a hermeneutics of indeterminacy', *Currents in Biblical Studies,* 7, pp. 181-193.

Douglas-Klotz, Neil (2000). 'Genesis Now: Midrashic Views of Bereshit Mysticism in Thomas and John'. Paper presented at the Society of Biblical Literature Annual Meeting in the Thomas Traditions Section, Nashville, TN, November 21, 2000.

Douglas-Klotz, Neil (2003). *The Genesis Meditations: A Shared Practice of Peace for Christians, Jews and Muslims.* Wheaton, IL: Quest Books.

DSM-III-R (1987). *Diagnostic and Statistical Manual of Mental Disorders,* third edition revised. Washington DC: American Psychiatric Association.

Dupré, L. (1996). 'Unio Mystica: The State and the Experience', in *Mystical Union in Judaism, Christianity, and Islam: An Ecumenical Dialogue,* eds. M. Idel and B. McGinn, 3-23. New York: Continuum.

Dionysius the Areopagite (1987). *The Complete Works* translation by Colm Luibheid. New York: Paulist Press.

Eflaki, Shemsu-d-din Ahmed (1976). *Legends of the Sufis: Selected Anecdotes.* Trans J. Redhouse. London: Theosophical Publishing House.

Einstein, Albert (1949). *The World as I See It.* Thinker's Library, London.

Elam, Jennifer with Ruthen, Linda (2005), *Raising our Voices, Telling our Stories,* (In preparation)

Ellison, Marvin M. (1996). *Erotic Justice: A Liberating Ethics of Sexuality.* Louisville KY: Westminster John Know Press.

Ernst, Carl W. (1997). *The Shambhala Guide to Sufism.* Boston and London: Shambhala.

Farber, Seth (1993), *Madness, Heresy and the Rumor of Angels,* Chicago. Ill: Open Court.

Feinstein, David and Krippner, Stanley (1988). *Personal Mythology. The Psychology of Your Emerging Self,* .London: Unwin Hyman

Fenton, J.Y. (1995). 'Mystical Experience as a Bridge for Cross-Cultural Philosophy of Religion: A Critique', in *Religious Pluralism and Truth: Essays on Cross-Cultural Philosophy of Religion,* ed. T. Dean, 189-204. Albany, NY: SUNY Press.

Ferrer, Jorge. N. (1998). 'Speak Now or Forever Hold Your Peace. An Essay Review of Ken Wilber's *The Marriage of Sense and Soul: Integrating Science and Religion'*, *The Journal of Transpersonal Psychology* **30** (1): 53–67.

Ferrer, Jorge. N. (2000a). 'Transpersonal Knowledge: A Participatory Approach to Transpersonal Phenomena', in *Transpersonal Knowing: Exploring the Horizon of Consciousness*, eds. T. Hart, P. Nelson, and K. Puhakka, 213-252. Albany, NY: SUNY Press.

Ferrer, Jorge. N. (2000b). 'The Perennial Philosophy Revisited', *The Journal of Transpersonal Psychology* **32** (1): 7–30.

Ferrer, Jorge. N. (2002). *Revisioning Transpersonal Theory: A Participatory Vision of Human Spirituality.* Albany, NY: SUNY Press.

Fisher, Andy (2002). *Radical Ecopsychology,* Albany: State University of New York Press.

Floyd, Malcolm (1993). 'The Trouble with Old Men'. Paper given at a research seminar at King Alfred's College

Fordham, M. (1986). *Jungian Psychotherapy,* London; Maresfield

Foucault, Michel (1980). *Power/Knowledge: Selected Interviews and Other Writings 1972-1977* ed. Colin Gordon. New York: Pantheon Books.

Fox, Matthew (1988). *The Coming of the Cosmic Christ.* San Francisco: Harper and Row.

Freedman, Wendy L. (2003). 'Measuring and Understanding the Universe', To be published in *Reviews of Modern Physics Colloquia.*

Freud, Sigmund (1915). 'The Unconscious', in *The Standard Edition of the Complete Psychological Works of Sigmund Freud* Translated and edited by James Strachey and others. Vol.14 p. 158-215, London: Hogarth Press (1960).

Gablik, Suzi (1991). *The Re-Enchantment of Art* Thames and Hudson. New York, New York.

Gebser, Jean (1953). *The Ever-Present Origin,* Athens, Ohio: Ohio University Press

Gibran, Kahlil (1991). *The Prophet* London: Pan Books

Gilbert, P. (1992). *Depression. The Evolution of Powerlessness.* Hove, UK: Psychology Press

Gooch, Stan (1972). *Total Man: Towards an Evolutionary Theory of Personality,* London: Allen Lane, Penguin Press.

Grey, Mary (1995). 'Till we have faces', included in Atkinson-Carter, Gloria (ed.) 2001, *In Being: The Winton Lectures 1979-2000,* Winchester: King Alfred's, Winchester pp. 61–8.

Gribbin, John (1995). *Schrödinger's Kittens and the Search for Reality,* London: Weidenfeld & Nicholson.

Griffiths, P.J. (1991). *An Apology for Apologetics: A Study in the Logic of Interreligious Dialogue.* New York: Orbis Books.

Habermas, J. (1992). *Postmetaphysical Thinking: Philosophical Essays,* trans. W. M.Hohenarten. Cambridge, MA: The MIT Press.

Halbfass, W. (1988). *India and Europe: An Essay in Understanding.* Albany, NY: SUNY Press.

Halbfass, W. (1991). *Tradition and Reflection: Explorations in Indian Thought.* Albany, NY: SUNY Press.

Haraway, Donna (1992). 'Ecce homo, Ain't (Ar'n't) I a woman, and Inappropriate/d Others: The Human in a Post-humanist landscape', in *Feminists Theorize the Political*, ed. Judith Butler and Joan Scott. New York: Routledge

Hartle, James (1991). 'The quantum mechanics of cosmology', in *Quantum Cosmology and Baby Universes* ed. S Coleman, P Hartle, T Piran and S Weinberg, Singapore: World Scientific.

Harvey, Andrew (1998). *The Essential Mystics*. Edison, New Jersey: Castle Books.

Heim, S.M. (1995). *Salvations: Truth and Difference in Religion*. Maryknoll, NY: Orbis Books.

Helman, Cecil G. (1994). *Culture, health and Illness*, London: Butterworth/ Heinemann

Heron, John (1998). *Sacred Science: Person-Centred Inquiry into the Spiritual and the Subtle*. Ross-on-Wye, Herefordshire: PCCS Books.

Heyward, Carter (2003). Seminar on Christology, Episcopal Divinity School, Boston, MA.

Hick, J. (1983). 'On Conflicting Religious Truth-Claims', *Religious Studies* **19**(4): 485–91.

Hick, J. (1992). *An Interpretation of Religion: Human Responses to the Transcendent*. New Haven: Yale University Press.

Hillman, James (1996). *The Soul's Code: In Search of Character and Calling*. New York: Random House.

Hirsch, Edward (1999). *How to Read a Poem*. London: Harcourt.

Hixon, Lex (2003). *The Heart of the Qur'an: An Introduction to Islamic Spirituality*. Wheaton, IL: Quest Books.

Hollenback, J.B. (1996). *Mysticism: Experience, Response, and Empowerment*. University Park, PA: Pennsylvania State University Press.

Holt, John (2003). *Anthropologists of the Mind* Asylum Magazine Autumn 2003

Idel, M. (1988). *Kabbalah: New Perspectives*. New Haven: Yale University Press.

Isham, Chris J & Butterfield, Jeremy N. (1998). 'A topos perspective on the Kochen-Specker theorem: I. Mathematical development', *Int. J. Theor. Phys.* **37**, 11, 2669–733.

James, William (1902). *The Varieties of Religious Experience*. New York. Longmans.

Jantzen, Grace M. (1998). *Becoming Divine: Toward a feminist philosophy of Religion*. Manchester: Manchester University press

John of the Cross (1979). *Poems of St. John of the Cross*, Translated by Roy Campbell. Collins, London: Fount Paperbacks

Johnson, Elizabeth A. (1993). *She who is: The Mystery of God in Feminist Discourse* New York: Crossroads

Julian of Norwich (1977). *Revelations of Divine Love of Juliana of Norwich* trans. Del Mastro, M.L. New York: Doubleday.

Jung, Carl G. (1956). *Symbols of Transformation*. Trans. R.F.C. Hull. London: Routledge & Kegan Paul.

Jung, Carl G. (1968). *Psychology and Alchemy*, Trans. R.F.C. Hull , 2nd Ed. London: Routledge and Kegan Paul.

Jung, Carl G. (1980). *Man and His Symbols,* ed. C.G. Jung, Marie-Louise von Franz, John Freeman and others, London: Pan Books

Karpman, Stephen (1968). 'Fairy Tales and Script Dream Analysis' in *Transactional Analysis Bulletin* **7**, April 1968.

Kassi, Norma (1996). 'A Legacy of Maldevelopment: Environmental Devastation in the Arctic', in *Defending Mother Earth: Native American Perspectives on Environmental Justice,* ed. Jace Weaver, Maryknoll, NY: Orbis Books.

Kasulis, T.P. (1988). 'Truth Words: The Basis of Kukai's Theory of Interpretation', In *Buddhist Hermeneutics,* ed. D.S. Lopez Jr., 257–72. Honolulu: University of Hawaii Press.

Katz, S.T. (ed.) (1978). *Mysticism and Philosophical Analysis.* New York: Oxford University Press.

King, Mike (2002). 'Against Scientific Magisterial Imperialism', *Network,* April 2002, pp. 2–7.

King, Mike (2003). 'Towards a Postsecular Society', *Sea of Faith Magazine ,* Spring 2003.

Koestler, Arthur (1975). *The Ghost in the Machine.* London: Pan Books.

Küng, H. (1991). *Global Responsibility: In Search for a New World Ethic.* New York: Crossroad.

Küng, H. and K-J. Kuschel (eds.) (1993). *A Global Ethic: The Declaration of the Parliament of the World Religions.* New York: Continuum.

Lawrence, T.E. (1962). *Seven Pillars of Wisdom.* Harmondsworth: Penguin Books.

Levy-Bruhl, Lucien (1985). *How Natives Think.* Princeton University Press.

Lewis, C.S. (1959). *The Lion, the Witch and the Wardrobe.* Harmondsworth: Puffin Books.

Lieberman, J.N. (1977). *Playfulness: Its Relation To Imagination and Creativity.* New York: Harcourt, Brace, Jovanovich.

Lorimer, David (ed.) (1998). *The Spirit of Science,* Edinburgh: Floris Books.

Mantin, Ruth (2002). *Thealogies in Process: The Role of Goddess-talk in Feminist Spirituality.* Unpublished PhD Thesis, May 2002, University of Southampton.

Maslow, Abraham H. (1962). *Towards a Psychology of Being,* Princeton New Jersey: D. van Nostrand Company Incorporated.

Maslow, Abraham H. (1964). *Religions, Values and Peak-Experiences.* Columbus, OH: Ohio State University Press.

Maslow, Abraham H. (1993). *The Farther Reaches of Human Nature.* New York: Penguin Arkana.

Maslow, Abraham H. (1996). *Future Visions: The Unpublished Papers of Abraham Maslow.* Ed. Edward Hoffman, Thousand Oaks, CA: Sage Publications.

Matte Blanco, Ignacio (1975). *The Unconscious as Infinite Sets.* London: Duckworth.

Matte Blanco, Ignacio (1988). *Thinking, Feeling and Being.* Routledge, London.

Maugham, W.S. (1963). *The Razor's Edge.* Harmondsworth: Penguin Books.

McGinn, B. (1996). 'Comments'. In *Mystical Union in Judaism, Christianity, and Islam: An Ecumenical Dialogue*, ed. M. Idel and B. McGinn, 185–93. New York: Continuum.

McNiff, Shaun (1992). *Art as Medicine. Creating a Therapy of the Imagination*. Boston, London: Shambhala.

Merchant, Caroline (1983). *The Death of Nature: Women, Ecology and the Scientific Revolution*. New York: HarperCollins.

Mookergee, Ajit (1998). *Ritual Art of India*. London: Thames and Hudson.

Moore, Thomas (1992). *Care of the Soul*. New York: Harper Collins.

Myers, I.B. (1993, original edition 1980). *Gifts Differing: understanding personality type*. Palo Alto, California: Consulting Psychologists Press.

Myers, I.B. and McCaulley, M.H. (1985). *Manual: a guide to the development and use of the Myers-Briggs Type Indicator* (2nd edition). Palo Alto, CA: Consulting Psychologists Press.

Nagel, Thomas (1986). *The View from Nowhere*. New York: Oxford University Press.

Nasr, Seyyed Hossain (1968). *Man and Nature: The Spiritual Crisis in Modern Man*. London: Unwin.

Nasr, Seyyed Hossein (1989). *Knowledge and the Sacred*, Albany, NY: State University of New York.

Nasr, Seyyed Hossein (1991). *Sufi Essays*. Albany: State University of New York Press.

Nhat Hanh, Thich (1987). *Interbeing*. Berkeley, Ca: Parallax Press.

Nicholas of Cusa (1997). *Nicholas of Cusa: Selected Spiritual Writings*, trans. Hugh Lawrence Bond. Mahwah, New Jersey: Paulist Press.

Okri, Ben (1996). *Birds of Heaven* London: Phoenix Paperback

Ormiston, Gayle L and Schrift, Alan D (eds.) (1990). *Transforming the Hermeneutic Context: From Nietzsche to Nancy*. Albany: State University of New York Press,

Panikkar, R. (1984). 'Religious Pluralism: The Metaphysical Challenge', in *Religious Pluralism*, ed. L.S. Rouner, 97–115. Notre Dame, IN: University of Notre Dame Press.

Panksepp, J. (1998). *Affective Neuroscience*. New York: Oxford University Press.

Peck, M Scott (1978). *The Road Less Travelled*. New York: Touchstone/Simon and Schuster.

Perry, John Weir (1974). *The Far Side of Madness*. Englewood Cliffs, NJ: Prentice-Hall.

Perry, John Weir (1999). *Trials of the Visionary Mind : Spiritual Emergency and the Renewal Process*. Albany: State University of New York Press.

Plato (1955). *The Republic*, trans. H.D.P. Lee. Harmondsworth: Penguin.

Primavesi, Anne (2000). *Sacred Gaia*. London and New York, Routledge.

Primavesi, Anne (2003). *Gaia's Gift: Earth, Ourselves and God after Copernicus*. London and New York, Routledge.

Primavesi, Anne (2004). *Making God Laugh: Human Arrogance and Ecological Humility*. Santa Rosa, Polebridge Press.

Ralph, R. and Kidder, K. (2000). *What Is Recovery? A Compendium of Recovery and Recovery-Related Instruments*. Cambridge MA: Human Services Research Institute.

Ravindra, Ravi (2000). *Science & the Sacred*. Chennai: Theosophical Publishing House.

Ray, Paul H. and Anderson, Sherry Ruth (2000). *The Cultural Creatives: How 50 Million People Are Changing the World*. New York: Harmony Books.

Reason, Peter and Rowan, John (eds.) (1981). *Human Inquiry: A Sourcebook of New Paradigm Research*. Chichester: John Wiley & Sons.

Redfearn, Joseph (1992). *The Exploding Self: The Creative and Destructive Nucleus of the Personality*. Wilmette, Illinois: Chiron Publications.

Reich, Wilhelm (1948). *The Function of the Orgasm*. New York: Simon & Schuster.

Reich, Wilhelm (1949). *Character Analysis*. New York: Farrar, Straus and Giroux.

Reich, Wilhelm (1983). *Children of the Future*: New York: Farrar, Straus and Giroux.

Richards, Mary C. (1973). *The Crossing Point*. Middletown, Conn: Wesleyan University Press.

Rogers, Carl (1976). *On Becoming a Person*. London: Constable.

Rowan, John (1981). 'A Dialectical Paradigm for Research', in Reason and Rowan, (1981) pp. 93–112. See above.

Rowan, John and P. Reason (1981). 'On Making Sense', in Reason and Rowan, (1981) pp. 113–37. See above

Rumi, Jalal al-Din (1898). *Selected Poems from the Divani Shamsi Tabriz* edited and translated by Reynold A. Nicholson. Cambridge: Cambridge University Press.

Russell, Robert John; Philip Clayton; Kirk Wegter-McNelly and John Polkinghorne (eds.) (2001). *Quantum Mechanics: Scientific Perspective on Divine Action*. Vatican City State: Vatican Observatory; and Berkely, Calif.: Center for Theology and Natural Sciences.

Ryle, Anthony (1995). *Cognitive Analytic Therapy*. Chichester: Wiley.

Sandoval, Chela (2000). *Methodology of the Oppressed*. Minnesota: University of Minnesota Press.

Schmimel, Annemarie (1975). *Mystical Dimensions of Islam*. Chapel Hill: University of North Carolina Press.

Schimmel, Annemarie (1992). *Islam: An Introduction*. Albany: State University of New York Press.

Schimmel, Annemarie (1994). *Deciphering the Signs of God: A Phenomenological Approach to Islam*. Albany: State University of New York Press.

Schore A.N. (1994). *Affective Regulation and the Origins of the Self*. New York: Lawrence Erlbaum Associates.

Sells, Michael (1996). *Early Islamic Mysticism: Sufi, Quran, Miraj, Poetic and Theosophical Writings*. New York: Paulist Press.

Sells, Michael (1999). *Approaching the Quran: The Early Revelations*. Ashland, OR: White Cloud Press.

Shlain, Leonard (1998). *The Alphabet Versus The Goddess*, London: Allen Lane, Penguin Press.

Shoham, Shloma Giora (2003). *Art, Crime and Madness*. Brighton: Sussex Academic Press.

Shrader-Frechette, Karen (2003). 'Ecology', in *A Companion to Environmental Philosophy*, ed. D. Jamieson, pp. 304–15. Oxford: Blackwell.

Sjørup, L. (1998). *Oneness: A Theology of Women's Religious Experience.* Leuven: Peeters.

Smith, H. (1989). *Beyond the Postmodern Mind,* 2d ed., updated and revised. Wheaton, IL: The Theosophical Publishing House.

Smith, H. (1994). 'Spiritual Personality Types: The Sacred Spectrum', in *In Quest of the Sacred: The Modern World in the Light of Tradition,* eds. S. H. Nasr and K. O'Brien, pp. 45-57. Oakton, VA: Foundation for Traditional Studies.

Soelle, Dorothee (2001). *The Silent Cry* trans. Barbara and Martin Rumscheidt. Minneapolis: Augsburg Fortress.

Spretnak, Charlene (1991). *States of Grace: the recovery of meaning in the postmodern age.* New York: HarperCollins.

Stebbins, R.A. (1992). *Amateurs, Professionals and Serious Leisure.* Montreal/Kingston: McGill-Queen's University Press.

Stoeber, M. (1994). *Theo-Monistic Mysticism: A Hindu-Christian Comparison.* New York: St. Martin's Press.

Tagore, Rabindranath (1921). *Lover's Gift and Crossing.* London, Macmillan.

Teasdale, John D. and Barnard, P.J. (1993). *Affect, Cognition and Change: Remodelling Depressive Thought.* Hove:Lawrence Erlbaum Associates.

Thalbourne M.A., Bartemucci L., Delin P.S., Fox B., and Nofi O. (1997). 'Transliminality: Its nature and correlates'. *The Journal of the American Society for Psychical Research* **91**: 305–31.

Thibaut, G. (trans.) (1904). *The Vedanta Sutras with the Commentary of Ramanuja. Sacred Books of the East, Vol. 48.* Oxford: Clarendon Press.

Tillman J.B. (1987). *Towards a Model of the Development of Musical Creativity: A Study of the Compositions of Children aged 3–11.* Unpublished PhD Thesis, University of London Institute of Education.

Torbert, William R. (1981a). 'Why Educational Research Has Been So Uneducational: The Case for a New Model of Social Science Based on Collaborative Inquiry', in Reason and Rowan (1981) pp. 141-151. See above

Torbert, William R. (1981b). 'Empirical, Behavioural, Theoretical and Attentional Skills Necessary for Collaborative Inquiry', in Reason and Rowan, (1981) pp. 437–46. See above.

Tubbs, Nigel (1998), 'What is Love's Work?', in *Women: A Cultural review* Vol **9**, No.1 October pp. 34–46.

Underhill, Evelyn (1911). *Mysticism: A Study in the Nature and Development of Man's Spiritual Consciousness.* London : Methuen

Underhill, Evelyn (1920). *The Essentials of Mysticism.* London, Dent and Sons.

Urantia Foundation (1955). *The Urantia Book.* Chicago: Urantia Foundation.

Varela, Francisco J., Thompson, Evan, and Rosch, Eleanor (1991). *The Embodied Mind: Cognitive Science and Human Experience.* Cambridge, MA: The MIT Press.

Vroom, H.M. (1989). *Religions and the Truth: Philosophical Reflections and Perspectives.* Grand Rapids, MI: William B. Eerdmans Publishing Company.

Wallas, Carl (1926). 'The art of thought', in Vernon, Philip E. (ed.) *Creativity,* Harmondsworth: Penguin pp. 91–7, 1970.

Ward, Hannah & Wild, Jennifer, (1995). *Guard the Chaos: Finding Meaning in Change.* London: Darton, Longman and Todd.

Wason, P. and Johnson-Laird, P. (1972). *Psychology of Reasoning: Structure and Content.* Harvard University Press, Cambridge, MA.

Watt, Douglas F. (1990). 'Higher cortical functions and the ego: the boundary between psychoanalysis, behavioral neurology and neuropsychology', *Psychoanalytic Psychology* 7(4), 487–527.

Watt, Douglas. F. and Pincus D.I. (2004). 'The neural substrates of consciousness: implications for clinical psychiatry', in: J. Panksepp, (ed.), *Textbook of Biological Psychiatry.* Hoboken, New Jersey: Wiley.

Wiggins, J.B. (1996). *In Praise of Religious Diversity.* New York: Routledge.

Wilber, Ken (1995). *Sex. Ecology and Spirituality: The Spirit of Evolution.* Boston, MA: Shambhala.

Wilensky, H.L. (1960). 'Work, Careers and social integration', *International Social Science Journal,* 12, 543–60.

Winnicott, D.W. (1987). *The Child, the Family and the Outside World.* Reading MA: Addison-Wesley Publishing.

Wittgenstein, Ludwig (1961). *Tractatus Logico-philosophicus.* London, Routledge & Kegan Paul

Wolfram von Eschenbach (1961). *Parzifal* trans. H.M. Mustard & C.E. Passage. New York: Random House

Zaehner, R.C. (1960/1994). *Hindu and Muslim Mysticism.* Rockport, MA: Oneworld Publications.

Index

Figures in bold type denote chapters or editorial sections

A Course in Miracles, 164, 165
Abi'l Khayr, 193
aboriginal. *See* indigenous peoples
Abram, David, 4, 93, 96, 145-6,
    178-9, **200-217**, 224
Acot, P, 223
action research, 185
affect. *See* emotion
Aldridge, David, 11
al-Hallaj, 194
amygdala, 98
anarchy, spiritual, 114
anawim, 234
Andrews, Lyn, 4, 106, **159-76**
angels, 95, 239
apophatic mysticism, 127, 140
Aristotle, 139
  logic in, 139
  Prior Analytics, 148
armor, in Reich, 197
arousal, 93
art, 41
  and medicine, 38
  creative source of, 51
  in Reich, 198
  outsider, brut, 47, 49
  pathological, 61
  recovery of, 47
  ritual, 42
art therapy, 39
Assagioli, Robert, 11, 15,
association in bilogic, 153

astronomy, perspective of, 206
attachment, 67, 71-7, 82, 84, 88
Attar, 192, 193
attention
  interpenetrating, 186
  valuing different forms of, 220
Augustine, 22, 141
awareness, integrating levels of,
  220
awe, 67, 71-2, 85, 97
  and also fear, 226

baby, no such thing as a, 95
Bache, Chris, 1
Bakan, J, 219
Ball S J, 9
Barnard, Philip, 68, 93, 94
Bateson, Gregory 179, 225-7
Bayat, Mojdeh, 192
beauty, 140
  of cosmos, 86
Beena, C, 124
being-and-the-world, 126
Belenky, Mary Field, 9
Berman, Morris, 36, 37
Best, Ron, 235
Bible, 91
big bang, 77-9, 171
bilogic, 105, 141 ff
  and association, 153
  context dependence of, 152
bi-logical structure, 133

birds, conference of (Attar), 192
Bistami, 190, 192
Blake, William, 84
bodhisattva, 194
body
  and attachment, 73
  awareness, 186, 188
  balance with mind, 37
  being in, 90
  body–mind split, 21, 205
  denied in mysticism, 229
  distortions in relation to, 22
  evolution, 211
  in ICS, 93, 99
  in mysticism, 57
  spiritual knowledge and, 108
  split from soul, 68
  world of, 210
body sensation, unity of, 198
Boman, Thorlief, 194
Bomford, Rodney, 4, 98, 105,
  129-42, 145, 148, 151-3, 177
both/and logic, 12, 98, 103, 146,
  158-9, 167
boundaries
  loss of, 98, 145
  loss of in psychosis, 97
  opening of in mysticism, 145
  personal, 26
  repairing, 34
  setting, 166
Boyce-Tillman, June, 3, 6-7, **8-33**, 69,
  105, 178
brain
  and ICS, 169
  attachment mechanisms, 73, 88
  damaged by trauma, 48, 82
  infant, 88
  left/right, 69
  limits to our comprehension of,
    212
  processing systems, 98
brain science inadequacies, 205
Brown, D P, 112
Bruteau, Beatrice, 22, 23

Bucky, Peter A, 51
Buddhism, 112-3, 118, 122, 177
  variety within, 118
Buswell, R E, 118, 122

call, 172
Campbell, Joseph, 85
capacities and spiritual ranking,
  119
Capra, Fritjof, 36, 37
caregivers, 77, 94-5, 101
challenge/nurture, 32
channelling, 167
chaos
  and creativity, 13
  as sacred, 64
  domestication of, 48
  in grieving, 15
  of de-integration, 7, 13
  temporary, 64
  through bilogic, 133
  transformation by story, 35
  versus order, 14
chaos theory, 79
child development, 100
childhood traumas, 75, 82
child-like wonder, 54
children
  as powerless group, 220
  brains damaged by abuse, 48, 82
  decide their future (Reich), 199
  educated as consumers, 7
  effect of rejection on, 52
  home education, 51
  one with nature, 198
  suppression in, 197
  testing of, 14
  violence done to by television,
    216
church, 5, 25, 62, 99, 141
Clarke, Chris, 12, 104, **143-58**, 164 ff,
  175-6
Clarke, Isabel, 3, 68, **90-102**, 106,
  145, 167

classification as a bilogical activity, 138

Clement of Alexandria, 140

Cloud of Unknowing, The, 133, 141

co-creation, 97, 107, 116

cocreators, humanity as, 235

cognition in Maslow, 196

cognitive subsystems, interacting. *See* interacting cognitive subsystems

Cohen, Leonard, 90

cold mechanisms, 78

collaborative research, 185

communication, 171

communion, 108, 120, 128, 157, 189, 211-2, 230-2
  with God, 229

community, 6, 13, 25 ff, 46, 48, 54, 188, 196, 213, 225, 235
  in interpretation, 182, 186-7
  in new paradigm research, 185, 186
  integral to any mysticism, 222
  more-than-human, 224

community of inquiry, 187

computer, dangers of, 216

condensation (psychological), 132

connected way of being, 157

consciousness, 246
  and integrity, 165
  and quantum theory, 174
  and spirit, 168
  as emergent, 78, 79
  aspects of, 168
  changes in, 106
  disembodied, 86
  duality in, 90
  ecological, 221, 229-31, 233
  eternal, 86
  genesis of, 77
  global, 89
  higher, 167
  heightened, 45, 50
  holistic level, 197
  in *The Urantia Book*, 168

integral, 167
loss of, 162, 163
ma'na, 192
modern, 35
overwhelming, in mysticism, 229
participatory, 37, 108
raising of, 171
rational, 128
refinement of, 2
scientific explanation, 143
stages of development, 192
states of, 2, 181, 191
structure of, 168
transcendent, 123
transcendent acts of, 22
view of unconscious from, 132

consciousness studies, 105

context-dependent logic, 152

contexts and logic, 105

contradiction, 15, 103, 105

conversation with the world, 158

Coomaraswamy, Ananda K, 45

Copernicus, 244
  jolt to experience, 205

Corbin, Henry, 183

corporate activity, 219

corporate body, 219

Cowling, Jamie, 49

crack, 93

creation
  and quantum theory, 174
  creativity, 13, 16, 246
  and de-integration, cyclical, 14
  and sense of self, 6
  as immune system of mind, 50
  aspects of within process, 174
  can influence the divine, 115
  democratising, 18
  in relation to matter, 175
  insights from, 7
  loss of through medication, 45
  manifests to heal, 38
  pain and, 41
  participatory, 171

self-managed, 42
suppression of and ill health, 47
creativity or Spirit, 127
Creedmoor Psychiatric Center, 38
Crick, Francis, 2
cyberspace, 201

daimon, 172
daimonic, 61
Dalai Lama, 110-3, 118, 124
Danchin, Lauren, 47
dark night, 61, 63
darkness, *see also* light
   and chaos, 14
   and creativity, 13
   and fear, 56
   as gift, 64
   defeating, 41
   dynamic with light, 115
   in mystical path, 61
   underworld, 13
   versus light, 16
Darwinism, 223
Dean, T, 124
death
   and religions impulse, 74
   and underworld, 16
   cycle of, 42, 84
Descartes, 13, 76, 104, 119-20, 125,
   205
desert, 40, 44, 55, 96
design, in universe, 82
Dewey, John, 11
Diagnostic Statistical Manual of Men-
tal Disorders, see DSM
different and same, interplay of,
   137
Dionysius the Pseudo-Areopagite,
   131, 141, 148
displacement (psychological), 132
distance, dissolving, 200
distributive law, 151
diversity, 13
divine, nature of, 115

dominant way of knowing.
   See subjugated way of knowing
Douglas-Klotz, Neil, 4, 177, **180-99**
dreams, 41, 63, 94, 96, 99, 100, 171
drugs, 45, 57
DSM, 172, 239
duality, wave-particle, 152
Dupré, L, 115, 116

earth
   animate, 213
   awareness of, 211
   community of life, 231
   connection with, 95
   distancing from it, 224
   our inability to live on, 92
   relationship with, 97, 102
   resources, 221
   roots of all in, 221
   sustaining the, 36
   violence to, 19, 27, 28, 87, 96, 101,
      see also ecological crisis
   web of interactions, 221
Eckhart, Meister, 141
ecological
   awareness. *See* consciousness,
      ecological
   crisis, 67, 81, 89, *see also* earth,
      violence to
   humility, 231
   interpretation of scripture, 188
ecology, 36, 80, 81, 178, 225
   age of, 222
   as basis for knowing, 222
   difficulty of framing laws in, 225
   origins of, 223
   uniting other sciences, 224
ecopsychology, 179
ecosystem, as tool for awareness,
   225
ecosystems, 36, 80
   objections to concept, 225
ecstasy, 97-8, 195
education, 11, 15, 95, 150, 164, 176,
   235

Eflaki, Shemsu, 192
ego, 36, 165, 172, 176, 184, 196, 246
Einstein, Albert, 51, 70, 72, 74, 154
either/or logic. *see* both/and logic
Elam, Jennifer, 3, 7, **51-66**, 106, 146, 178
Ellison, Marvin M, 26
emancipation, 111, 125, 176
emancipatory power, 122-3, 127, 176
embodied/disembodied, 21
emergence, 68, 71, 76, 78-9, 84, 86, 175
emotion
  affective mystics, 115
  and curiosity, 72
  and unconscious, 129, 137
  core of spirituality, 67
  displacement of, 132
  in belief systems, 73
  in ICS model, 93-4
  in religion, 71
  in religions searching, 68
  infinitising, 132
  right hemisphere, 74
  side of being, 177
  states, 87
empathy, 84, 87-8, 161, 166, 216
  somatic, 215
emptiness, 48, 112-3, 118, 127
enaction, 109, 112-6, 122-3, 157
enchantment, 36-7, 160
Enlightenment, the, 10,13, 18-21, 25-6, 139
enlightenment, 112, 124
environment and mysticism, 224
epistemology, 61, 84, 109, 126, 187
Ernst, Carl W, 181, 190, 223
estrangement, 157, 208
eternal consciousness, 86
ethical neutrality, 80
ethical sensibility, 214
ethics
  and bodily interaction, 215
  and religious common ground, 114
  based on feeling, 214
  grounded in primary reality, 217
  in science, 80, 81, 82
  relativism, 82
  rooted in senses, 214
evil
  as good not realised, 87
  dualistic category, 234
  fear not source of, 87
  problem of, 22
Evolutionary Deity, 168
excitement/relaxation, 30
experiential world, 146
externalities, corporate, 219, 223
extrovert, 17

faith, value of, 83
fall, doctrine of, 210
Farber, Seth, 50
Feinstein, David, 43
feminism, 14
Fenton, J Y, 112
Ferrer, Jorge, 1, 4, 104, 106, **107-28**, 149, 169, 176, 177
fertilisation of cosmos, self, 170
Fisher, Andy, 92, 179
Floyd, Malcolm, 25
forces, organising, 84
Fordham, M, 15
forest, 85, 96, 158
form, ideal in Plato, 139
forms, Platonic, 202
Foucault, Michel, 5, 9
Fox, Matthew, 7
fragmentation
  in our lives, 213
Freedman, Wendy, 1
freedom, 1, 25-6, 46, 135, 176
  spiritual, 125
Freud, Sigmund, 5, 75, 129, 131-2, 197
frogs, 90
fundamentalism, 69-70, 75-6, 101, 208, 234

Gablik, Suzi, 36, 37
Gaia theory, 225
Galileo, 205
Gandhi, 236
gathered meeting, 58
Gebser, Jean, 167
generalisation, law of, 137
genetic, 204, 206
    need for connection, 82
Gibran, Kahlil, 70
gift from God, 232
Gilbert, P, 100
God
    abandoned by, 75, 88
    and ecological awareness, 227
    and the Unconscious, 141
    anthropomorphic, 76
    as Mother, 228
    coincidence of opposites in, 131
    male models of, 228, 230
    mind of, 161
    paradox of trinity, 142
    presence of, 55, 60
Gooch, Stan, 8
goodness, Platonic, 140
gradations, spiritual, 117
grail, 91, 92
Gregory of Nyssa, 140
Grey, Mary, 16, 235
Gribbin, John, 174
grieving, 15
Griffiths, P J, 120
ground of being, 67, 76
grounded theory, 165
growth, 32

Habermas, J, 128
hadith, 190
Haeckel, 223-4, 225
hal, 181, 191-2, 195, 197
Halbfass, W, 117, 118
Haraway, Donna, 12
Hartle, James, 155
Harvey, Andrew, 222, 224, 228-33
Hayward Gallery, 49

healing, 199
    and ritual, 55
    capacity, 6
    divine process, 34
    experience of, 56
    holistic model, 45
    lack of cultural context for, 37
    reached through chaos, 7
    shamanic, 110
    spiritual, 59
healing place, a, 54
heaven, Christian, 203
Hebrew traditions, 178, 180, 233
Heidegger, Martin, 104
Heim, S M, 120
heliocentrism, 205
Helman, Cecil G, 11
hemisphere, right/left, 74
hermeneutics, 72, 120, 178, 180, 184,
    186-7
    of indeterminacy, 181
Heron, John, 115
Heyward, Carter, 11
Hick, J, 120
hierarchical integration, 197
hierarchy
    alternative to, 103
    and overcoming realism, 104
    critiques of, 5
    inclusive, 120
    mystical, 119
    natural, 84
    of attributes of divine, 127
    of emergence, 68, 71
    of mystical stations, 193
    of needs, 195
    of ranking traditions, 107
    of spiritual gradations, 117, 121-3
    of visions, 128
    of ways of knowing, 2
    pre-established, 119
    primate, 100
    social, 93
    study of, as liberation, 6

versus transformative approach, 165

Hildegard of Bingen, 19, 33, 166
Hillman, James, 39, 172
Hirsch, Edward, 200
Hixon, Lex, 184
holistic explanations, objections to, 225
Hollenback, J B, 112
Holt, John, 3, 6, 7, **34-50**, 178
Holy Spirit, 57, 142, 166
holy, the, assoc. with fear, 230
human tradition of separateness, 225
humility
  ecological, 231
  urgency of, 233

Iblis, 195
Ibn-'Arabi, 166
ICS, *see* interacting cognitive sub-systems
ideas, origin of, 161
Idel, M, 115
identity, 25
  in the unconscious, 133
ideologies, 85, 101
immanent, 123
immune system, 6, 37, 49-50
implicational subsystem, 93
incarnate, 22-3
indigenous peoples, 5, 14, 69, 149, 178, 208, 234
  named at Rio conference, 220
individual
  and community, 6, 25
  and society, 9, 11, 25
  creation or gift, 10
  self, 26
individuality, 94, 191
inductive versus deductive, 226
infinitising of feeling, 132
integration, 14-6, 36, 60, 159, 175
  of self, 168

integrity, 6, 97, 197, 207, 228
  rather than consciousness, 165
interacting cognitive subsystems, 68-9, 93, 104, 167, 169
  emotion in, 93-4
  subsystems not symmetrical, 146
inter-being with nature, 222
interdependence, 81, 222, 225, 231
  with animals, 96
  of all organisms, 223
interpretation
  levels of, 188
intersubjective, 185, 187
introvert, 17
intuitive
  and rational, 18, 40, 59
  as grasping truth, 147
  aspects of church, 19
  capacities, 60
  dark as the, 61
  eyes, 144
  marginalised, 10
  oppression of, 19
Isham, Chris J, 155
Islamic traditions, 178, 180-183, 186-187, 190

James, William, 99
Jantzen, Grace M, 22
Jesus
  paradox with, 98
  versus Constantine, 101
jewellery and the self, 101
John of the Cross, 43, 44
Johnson, Elizabeth, 23, 141, 149
Julian of Norwich, 19, 53, 59
Jung, Carl G, 9, 12, 15, 47, 63, 96, 99, 141

kabbala, 115, 182, 184
Kandinsky, Wassily, 41
Kant, Emmanuel, 104
Karpman, Stephen, 136
Kassi, Norma, 1
Kasulis, T P, 118

Katz, S T, 112
Kemp, Margery, 19
King, Mike, 235
kingdom, 176
Klein, Melanie, 140
knowing
    as more than concepts, 232
    false, 123
    related to creation, 176
    spiritual. *See* spiritual knowing
Koestler, Arthur, 34, 40
kundalini, 57
Küng, H, 114

landscape, 62, 68, 96, 99, 144, 188,
    201
language
    apophatic, 127
    developing, 49
    developing a, 61
    finding, through ecology, 231
    in Quran, 184
    logic and, 147
    meeting Jewish-Islamic
        mysticism, 187
    mystical, inherent, 35
    of mysticism, 59, 147
    pointing without describing, 153
    Semitic, 181-2
    symbolic, 48
        self-realization through, 34
law of generalisation, 152
Lawrence, T E, 160-1, 163, 165-6
leisure, 9, 21, 28, 96
lens, 63, 64, 198
levels
    of knowing, 63
    of the person, 123
Levy-Bruhl, Lucien, 149
Lewis, C S, 91
liberation, 14, 107, 110-113, 116,
    125-126
    theology, 14
    spiritual, 124
Lieberman, J N, 11

light
    and dark, 61
    Christian concept, 13
    experience of, 59
    inner, in Marxism, 233
    John of the Cross, 43
    linked with truth, 14
    symbol of creation, 171
limbic system, 98
limen, 97
logic, 3-4, 8, 9, 12, 94, 103, 105,
    129-42, 148-56, 167, 206, 234
    and histories interpretation, 154
    and illogic, 99
    and language, 147
    and truth, 144
    Aristotelian, 139
    classical and symmetric, 105
    context-dependent, 152, 158
    enlarged, science of, 146, 153, 156
    history of, 148
    in practice, 149
    mathematical, 149
    prescriptive or descriptive, 149
    proscriptive, 114
    quantum, 151
    sub-dominant mode of, 98
Logos, 183
Lorimer, David, 173
love
    and body-image, 24
    and creativity, 173
    and fear, 163, 173
    as Buraq, 195
    bliss, 163
    by God as unlimited, 231
    ethic of, 124
    for God, 115
    from knowledge of connection,
        222
    in childhood, 82
    key to integration, 199
    mystical, 166
    of God by mystic, 229
    overcoming fear, 236

prelude to mysticism, 160
unconditional, 165
valley of, 193
wise and active, 229
Lovelock, 225

magical thinking, 76-7, 86
Mansur al Hallaj, Hussan ibn, 191
Mantin, Ruth, 23
maqam, 181, 191, 195, 197
Marton, Janos, 38, 45, 48
masks of God, 84-5
Maslow, Abraham, 11, 178, 181,
    195-9
mathematical worlds, 202
Matte Blanco, Ignacio, 105, 129-33,
    136-7, 141, 148, 151-2
maturity, 17
Maugham, W S, 172
McFague, 227
McGinn, B, 115
McNiff, Shaun, 46
meaning in quantum mechanics,
    156
mechanical universe, 78, 88
meditation, variety of, 112
Merchant, Caroline, 5
Merleau-Ponty, 104
midrash, 181, 182, 184, 186, 187
mind, 168
    and quantum context, 174
    double, 173, *see also* ICS
    in cosmos, 168
miracles, 99, 181
    absence of, 86
molecular biology and reality, 204
Mookergee, Ajit, 42
Moore, Thomas, 61
more-than-human community, 211,
    224
mother, wisdom of, 222
Murray, Henry, 43
Myers, I B, 9, 12
Mystery, 4, 108, 110, 114, 116, 123,
    127-8, 169, 176, 242

mystic, 7, 24, 43, 58, 115, 119, 163-4,
    166, 182, 196, 229,
    see also mysticism
mystical state, 181, 191, 196, 198, *see
    also* hal
mystical station, 181, 191, *See also*
    maqam
mysticism
    and attachment, 67
    and/of everyday life, 181, 191
    and ICS, 69
    and psychoanalysis, 4
    and self awareness, 106
    and the environment, 224
    apophatic, 140
    challenging conventional truth,
        148
    classical, 53
    contemporary, 54
    Einstein on, 51
    everyday life as, 199
    feminist, 53
    help in navigating, 52
    in childhood, 52
    incarnate, 22
    Jewish, 181
    Middle Eastern, 4, 177
    need for community belonging
        in, 222
    of daily life, 190
    opposed by traditional God
        image, 71
    repressed, 7
    source of creative, 51
    stories of experience, 52
    stripping away male discourse,
        23
    surveys of, 52
    symmetric logic and, 140
    the journey, 61
    Underhill, 61, 229
    underlying religion, 55
myth, 35, 42-3, 83-5, 99, 100-1

Nagel, Thomas, 121, 243

Nasr, Sayyed Hossein, 1, 2, 181, 183, 191, 193-4
nature, 13, 194
  amphibious, 90
  as hierarchy, 71
  as manuscript, 188
  as object, 67
  children at one with, 198
  for Haeckel, 225
  framework for, 80
  in Islamic cosmology, 183
  in Taoism, 48
  interbeing with, 222, 233
  intermeshing systems in, 36
  intimacy with, 231
  knowledge of, 232
  older Western view of, 36
  sacralisation of, 184
  sense of dependence on, 220
  understanding, 36
  violent approach to, 228
negative spiritual states, 195
neocortex, 98
neurological sciences
  claims of, 204
neuroscience, 71, 73, 79, 98
  affective, 82
New Age, 19, 25
news, onslaught of hype in, 207
Nhat Hanh, Thich, 232-3
Nicholas of Cusa, 131, 141, 147-8

observer, 19, 37, 46, 136
Okri, Ben, 35, 48
ontology, 71, 84, 86, 109
openness, in interpretation, 187
opposites, coincidence of in God, 131
opposition, rule of, 130
oracy, oral culture, 28-9
orgasm, in Reich, 197
orgone, 198
Origen, 140
Ormiston, Gale, 184

'other', the; otherness, 11, 13, 27, 193, 201

pain, 56
Panikkar, R, 121, 125
Panksepp, J, 73
pantheism, 85
paradigm, new, in social science, 185
paradox, 68, 73, 92, 98-9, 106, 195
  in mystical language, 147
  of mystical in Wittgenstein, 226
  of trinity, 142
participation, 4, 6, 37, 104, 107-10, 113-22, 125-7, 171, 188, 201
  metaphysics of, 157
Parzifal, 91, 92
pattern that connects, 226
Paul, St., 68
peak experiences, 195, 196, 197
Peck, M Scott, 172
perceptual world, 203
perennialism, 107, 110, 111, 128
Perry, John Weir, 64
personality
  as structure, 167
  structure of consciousness, 172
Plato, 22, 68, 92, 139, 142, 202
  and logic, 139, 148
  cave, 92
Plotinus, 140, 229
pluralism, **103-6**, 120, 121, 126, 128
  and universalism, 108
  dynamic with universalism, 128
  linked with universalism, 169
  of contextual truth, 147
plurality, 103, 110-1, 124-8
  of spiritual ultimates, 126
Plutarch, 218, 220, 221, 222
polarity, 10-1, 22, 26-7, 31
pornography, 22
powerless groups (Rio), 220
practice, transformation of, 177
presence, 43, 53-6, 59-60, 90, 110, 158, 168, 213, 216

of God, 55, 60
of world, 145
spiritual knowing and, 109
Primavesi, Anne, 4, 178, **218-33**
Prinzhorn, 47, 49
process, 28
creative, 39
of creativity, 13
sacred only within, 23
process/product, 6, 27-8, 30
progress, 14, 27, 81, 92, 96, 236
propositional subsystem, 68, 93, 145
propositions, 134-5, 144, 149, 155
compound, 150
definition of, 130
in quantum logic, 151
psychology, **67-69**
psychosis, 37, 43, 45, 57, 64, 94, 97, 99, 133, 172, 238
public, 17-8, 24, 28, 46, 227, 229
public/private, 17
Pythagoras, 202
Pythes, 218

Quaker, 51, 56, 58-60, 65
quantum logic, 151
and context dependent logic, 155
quantum mechanics/theory
and creation, 174
history interpretation of , 154
incompleteness of, 156
origins of, 154
Quran, 181-4, 187-8, 195, 245
example of exposition, 188
Qushayri, 194

rain, drinking the, 217
Ralph, R, 235
rational mind in Plato, 202
rationalism, 13, 21
rationality, 144
and intuitive, 18, 40, 59, 60, 69, 144
and mystical, 68

challenging the, 20
combining with mystical, 146
destruction of, 133
split from experience, 146,
    see also Enlightenment, the
Ravindra, Ravi, 236, 244
Ray, Paul H, 235, 244
realism, 3, 4, 104, 158
reality
and perennialism, 110
multiplicity of, 111
no given uiltimate, 113
Reason, Peter, 180, 185
reciprocal roles, 96
reciprocation, and emtional attraction, 135
Redfearn, Joseph, 15, 244
reenchantment. *see* enchantment
reflection, rule of, 130
Reich,Wilhelm, 178, 181, 195, 197, 198, 199
relatedness, 76, 93, 94
relational subsystem, 68, 93, 145
relationship and the self, 100
relativism and post modernism, 104
religion
a sham, 75
a theological Rorschach, 75
and science, connecting, 70
and state power, 101
antagonism with science, 72
attempts to grasp the infinite, 85
common element in, 110
disparages sensory reality, 213
exclusivist, 120
is it a knowing, 68
multicentered, 121
open, 29
sciences of, 183
truth claims in, 116
religions, common ground, 83
religiosity, evidence of in public life, 227
religious theory, chaos in, 107

religious traditions, 65
representation, 28
rescue schema, 136
research cycles, 187
rhythm, 30
Richard of Saint Victor, 115
Richards, Mary C, 61, 62
Rio, UN conference in, 220
rite of passage, 42, 45
ritual, 10, 25, 42
    of hunters, 96
Rogers, Carl, 11
Rorschach, 74, 75
rotation, law of, 136
Rowan, John, 180, 185, 187
Rumi, 147, 192, 195
Russell, Robert, 236
Ruusbroec, Jan van, 115
Ryle, Anthony, 96

sacred
    and profane, 191
    in Bateson, 226
    in the ordinary, 181, 196
    inseparable from whole, 226
    sense of, 108
    wholeness, 226
    world as, 232
salvation, 177
same and different, interplay of,
    137
Sandoval, Chela, 12
satyagraha, 236
Schmimel, Annemarie, 182-3
Schore, A N, 74
science
    disparagement of senses by, 213
    ecological, 221
    ethical value of, 81
    extending foundations of, 144
    limits of, 143
    nature of, 144
    ontology for, 78, 84
scientism, 3, 71-2

self, 34
    correspondences with universe,
        229
    creation of, 100
    deeper, 43
    dynamics of, 128
    dysfunction of, 3
    fractured, 34
    images of, 75
    in ICS, 93
    in Reich, 198
    integrated, 168
    interpenetrates with world, 111
    journey to integration, 14
    model of, 12
    no-self, 118
    relationships of, 7
    relationships within, 11
    revelation of, 110
    routes to, 43
    self integration, 42
    transformation of, 109
    transformed with world, 109, 122
self-actualisation, 195
self-centeredness, 110, 111, 122
self-consciousness, 90
self-expression of God, 115
selflessness, 122
self-organisation, 79
self-realisation, 34, 37
Sells, Michael, 183-4, 191-2, 194
Semitic languages, 181
sex, 99, 138
shaman, 35, 37
shame, 123
Shlain, Leonard, 69, 245
Shoham, Shloma Giora, 41, 245
Shrader-Frechette, Karen, 225, 232
Sjørup, L, 227, 230
skilful means, 108, 122
Smith, Huston, 121, 124-5
social context, **4-7**
society, 12, 17, 52, 173
Soelle, Dorothee, 147-8
somatic empathy, 216

soul, 25, 44, 168, 172
  evolution of, 159, 176
  evolutionary, 172
  going out to Soul, 183
  in Plato, 203
  in Underhill, 229
  loss of, 38
  split from body, 21, 68
Spinoza, 141
spirit
  creative, 168
  pre-personal, 172, 174
spiritual
  as values and perceiving, 144
  disciplines in Maslow, 196
  gradations, 117
  knowing, 2, 5, 61, 107-10, 113, 122
  path, 50, 58, **177-179**
  quality of experience, 97
spiritual reality, 60, 120, 122
  structures of, 163
spirituality, 107
Spretnak, Charlene, 104
state reduction, 175
state, mystical. *see* mystical state
station, mystical. *see* mystical station
Stebbins, R A, 9, 245
Stoeber, M, 119, 245
story, 7, 19-20, 22, 35, 48, 52, 54-5, 61, 63, 65, 74, 98-100, 159, 161, 177, 182, 184
strata of mental life, 132
structure, bilogical, 133
subjugated ways of knowing, 3, 5-6, 9-12, 18, 23-24, 26, 33, 177, 178
suffering, creative response to, 41
Sufi, 112, 181-2, 184, 190-8
supernatural, 99, 100
symbolic sight, 160, 166, 167
symmetric logic, 105, 129-32, 134-5, 137-9, 142, 152, 153
  not predetermined, 134
systems theory, 36

ta'wil, 180-8
Tagore, Rabindranath, 231-2
tawhid, 191, 195, 198
Teasdale, John, 68, 93
Teicher, Martin, 48
Teresa of Avila, 53, 59, 115
territoriality, 87
Thalbourne, M A, 94
theology, ecological awareness and, 227
theurgical mysticism, 115
Thibaut, G, 117
Tillich, Paul, 76, 83, 84
timelessness, 131
Torbert, William, 185-6
trance, 1, 161, 162, 167
transcendence, 22, 25, 70, 123, 208, 215
  allure of, 201
  in Maslow, 197
  internal, of every being, 226
  of mathematics, 202
  primacy over the earthly, 203
transference, 95
transfiguration, 91
transformation, 7, 50, 122
  of God, 115
  of self and world, 109
translation of Quran, 183
transliminal, 94-6, 99-101
tribalism, 42, 87-8, 96, 208
truth
  as aletheia, 148
  as attunement, 124
  absolute, 104
  contextual, 147
  correspondence theory of, 145
  one, 127
  Platonic, 140
  propositional, 146
Tubbs, Nigel, 15

ultimate concerns, 83

unconscious, 13, 62-3, 105, 129-42,
153
logic and, 131
Underhill, Evelyn, 61, 229, 230, 232,
246
understanding, 1, 4, 34-9, 41, 43, 45,
48, 50, 54, 58, 64, 65, 69
unitive in Maslow, 196
unity, 13, 191
mind and nature, 226
tawhid, 191, 198
universalism, 127-8
and pluralism, 108
universe
early, 79
mechanical, 88
self-fertilisation of, 170
Urantia Foundation, 164, 168, 171,
173

value
and quantum theory, 156
cognitive isomorphism with, 197
in ICS, 93
signals, 88
values
and images of caregivers, 77
emotion as origin of, 71
Enlightenment, 10
value-systems, 8-9
subjugated, 6
variety of, 11
Varela, Francisco J, 109, 114
virtual worlds, 215
visionary, 10, 19-20, 52, 61, 65, 110,
114, 116, 120, 181, 191, 199, 206
vocation, 172
Vroom, H M, 121, 124

Wallas, Carl, 13
Ward, Hanah, 14, 15

Wason, P, 149
Watt, Douglas, 3, 67, 68, **70-89**, 177,
179
wave-particle duality, 152
ways of knowing, intuitive, 159
ways of knowing, subjugated. *See*
subjugated ways of knowing
wealth, desire for, 219
web, 79-80, 157, 162, 211, 223
of connection/relations, 95
Whitehead, A N, 149
whole, sense of, in mysticism, 222
Wiggins, J B, 125
Wilber, Ken, 2, 112, 118, 119
Wilensky, H L, 9
Winnicott, D W, 95
wisdom, 23, 92
wisdom of the Mother, 222
Wittgenstein, 144, 226, 227, 247
Wolfram von Eschenbach, 91, 92,
247
work, 12, 28, 52
based on product, 29
manual, 21
world
another, 91, 92, 93, 200
of the body, 210
perceptual, 203
small and vast, of physics, 203
worlds
common source of all, 209
mathematical, 202
proliferation of, 206
scientific, not coherent, 205
Writers Digest, 159

yearning, 97

Zaehner, R C, 117
zen, 181
paradox in, 98

Journal of Consciousness Studies

www.imprint-academic.com/jcs

Journal
Consciousn
Stud

Volume 11, No.

controversies in science and the humanities

Trusting
the Subject?
Part 2

Journal of
Consciousness
Studies

Volume 12, No.3 (2005)

controversies in science and the humanities

Classical Hindu
psychology

Science, Consciousness & Ultimate Reality

Edited by David Lorimer

Science, Consciousness & Ultimate Reality

Edited by David Lorimer
250 pp., £14.95/$29.90, 0907845797 (pbk.)

Essays at the interface of science, religion and consciousness studies. Contributors include Denis Alexander, Bernard Carr, Chris Clarke, Guy Claxton, Peter Fenwick, David Fontana, John Habgood, Mary Midgley, Ravi Ravindra, Alan Torrance and Keith Ward.

'Vital.' *Science and Theology News*

Cognitive Models and Spiritual Maps

Edited by Jensine Andresen and Robert Forman

Cognitive Models and Spiritual Maps

Jensine Andresen/Robert Forman, ed.
288 pp., £14.95/$29.90, 0907845134 (pbk.)

This book throws down a challenge to the field of religious studies, offering a multidisciplinary approach — including developmental psychology, neuropsychology, philosophy of mind and anthropology.

'A thoroughly gripping read . . . I cannot possibly do justice to the complexity and sophistication of the positions on offer.' *Human Nature Review*

Shadow, Self, Spirit
essays in transpersonal psychology

Michael Daniels

Shadow, Self, Spirit

Michael Daniels
300 pp., £17.95/$34.90, 1845400224 (pbk.)

Transpersonal Psychology concerns the study of those states and processes in which people experience a deeper sense of who they are, or a greater sense of connectedness with others, with nature, or the spiritual dimension. Michael Daniels teaches the subject as part of the psychology curriculum, and this book brings together the fruits of his studies. It will be of special value to students, and its accessible style will appeal also to all who are interested in the spiritual dimension of human experience.